CRITICAL OPINIONS of the FIRST EDITION.

'This seems a very useful little volume, both as an Elementary School Text-Book and as suitable for the use of candidates preparing for matriculation, preliminary scientific, and other examinations. It is well put together, not too advanced for the end in view, and the exercises and illustrations are of a clear and pertinent character.'

LANCET.

'This little work is, as it claims to be, well suited for beginners, a very small knowledge of simple arithmetical rules and a smattering of algebra only being required to work out tolerably difficult problems. The treatment of the subject is novel, motion being considered before rest or equilibrium. Numerous examples are given throughout the book to be worked out by the student, and questions given at the Preliminary Scientific Examinations of the University of London for the last few years render the book very useful to the student.'

MIDLAND COUNTIES HERALD.

'In the study of Mechanics, as with Arithmetic, the principles are too frequently left unmastered, and a carefully set problem proves a stumbling-block, not generally from ignorance of formulæ, but from inability to apply the formula to the case in question. Preparation for examination in Elementary Mechanics requires a thorough grounding in the principles, and at least as much exercise in their application as is found necessary in Arithmetic and Algebra. Mr. MAGNUS's book seems admirably adapted to furnish these two requirements. . . . . . The order is scientifically logical, and appears to possess advantages in teaching. The principles are lucidly explained and illustrated; examples are worked out, and each lesson is followed by a series of graduated questions, requiring a thorough grasp of principles for their solution. An appendix gives a number of examination papers set at various public institutions.'

SCHOOLMASTER.

'Mr. MAGNUS's little work on *Elementary Mechanics* is a great improvement upon the few similar ones already in use, and will help many teachers and private students out of great difficulty. The importance of the subject is so generally recognised that almost every educated man is expected to know something of its leading principles; and there is scarcely a public examination held in which it is not either a necessary or an optional subject. Yet, strange to say, it has hitherto been no easy matter to find a text-book at once simple enough to meet the requirements of the beginner, and at the same time full and complete enough to serve as a basis for more advanced reading. Henceforth, however, it will only be necessary to name the one before us, and the difficulty will be at an end. Following the example recently set by the highest authorities on the subject, the Author has departed consider-

ably from the usual arrangement of the topics discussed. Kinematics, or the consideration of some of the simplest principles of motion, without regard to the quantity or quality of the matter moved, comes first. This is followed by the laws of Dynamics, under which head is considered the matter that is set in motion, and the cause or force producing it. The third and last part embraces all the problems connected with bodies at rest, which are discussed under the head of Statics, or the science of equilibrium. In addition to these subjects there is a brief but clear exposition of the doctrine of Energy, which affords a capital view of the connexion between the science of mechanics and other branches of physics. All the important propositions are accompanied and illustrated by numerical examples worked out in the text; and to each section are added progressively arranged exercises, which the student will find excellent tests of his knowledge of the principles laid down. To render the work more especially serviceable to candidates for the matriculation and preliminary scientific examinations of the University of London, all questions set during the last few years have been classified and appended to the several chapters to which they correspond.' LEEDS MERCURY.

'The style is lucid, the solved exercises carefully chosen, the work compact. . . . . An intelligent boy ought in a few months to be able to make himself master of the greater portion of this small book which Mr. MAGNUS has aimed at making sufficiently elementary to be placed in the hands of a beginner. What we consider to be higher praise is that we believe it to contain nothing that the student will have to unlearn in a subsequent portion of his career. We can recommend it as a trustworthy introduction to more advanced text-books.' NATURE.

Mr. MAGNUS'S book is admirably written from first to last, and shows remarkable powers of organising knowledge so as to adapt it to the gradually developing mind of the learner, and this shews the Author in his true colours as a thoroughly practical teacher. It is evident that the book is not made up of mere proposals or suggestions of what might be done, but it is a record of what has been done. It is plain that all these lessons have been given, that they are the results of experience; and in this consists much of their value. . . . Mr. MAGNUS realises the idea that a definition should not only explain the real meaning, but should be so worded that it could not possibly be taken to mean anything else, and this will save much confusion and misunderstanding on the part of the pupil. The definitions are in fact brief essays; and thus the pupil, instead of learning the mere words by heart, will, after reading them, become imbued with their spirit, will be interested in their application; and when this point is reached, half the difficulty of the study will have been removed.' JEWISH CHRONICLE.

LESSONS

IN

ELEMENTARY MECHANICS.

INTRODUCTORY to the STUDY of PHYSICAL SCIENCE.

DESIGNED FOR THE USE OF SCHOOLS, ACADEMIES, AND SCIENTIFIC INSTITUTIONS.

WITH NUMEROUS EXERCISES.

BY

PHILIP MAGNUS, B.Sc., B.A.,

LIFE-GOVERNOR OF UNIVERSITY COLLEGE, LONDON.

WITH EMENDATIONS AND INTRODUCTION

BY

PROF. DEVOLSON WOOD,

OF STEVENS INSTITUTE OF TECHNOLOGY.

SECOND EDITION, REVISED.

NEW YORK:

JOHN WILEY & SONS,

15 ASTOR PLACE.

1876.

John F. Trow & Son,
Printers and Stereotypers,
205–213 *East* 12*th St.*,
NEW YORK.

PREFACE.

ELEMENTARY works on scientific subjects are very desirable. It is, however, too often the case that in attempting to produce them they are really made more difficult than the more advanced works, and in too many instances *special* cases are treated as if they were *general*, and erroneous impressions are made upon the mind of the learner. Better not teach a subject than teach errors. It costs more to unlearn what one supposed to be true than to learn correctly the most abstruse principles. Mechanics has had its share of erroneous instruction in *elementary works;* but we believe that this one is truly elementary, logically correct, and well adapted to meet the wants of the young on this subject.

The article on "*Impulsive Forces*" has been re-written for the American edition, and so changed from the original as to conform with the writer's views of the subject, and, as we think, made it to conform more nearly with the views of the Author as expressed in other parts of the work.

My system of teaching agrees with that of the Author in regard to presenting some of the principles of Dynamics very early in the course, and I have adopted the plan in my larger work on Mechanics, which is now in press. A student gains a clearer and more general idea of the action of forces by considering their effect upon moving bodies than by considering them in Equilibrium. The Author has carried this part of the subject considerably beyond what is usually attempted in elementary works.

It is unnecessary to decide which is the more useful, Statics or Dynamics. They are equally indispensable. Both have numerous applications in the Arts and Sciences, and should receive their appropriate share of study.

The Author is direct and explicit in his statement of principles. The numerous and well-selected "Exercises" and problems are a grand feature of the work. The student who masters them cannot fail to have a good knowledge of the subject, and be able to apply its principles whenever required in practical life.

DE VOLSON WOOD.

HOBOKEN, *July*, 1876.

PREFACE.

In these Lessons, which are intended for the use of those who have had no previous instruction in the subject, I have endeavoured to bring into prominence the leading principles of Mechanics, and to exemplify them by simple illustrations; and with the view of showing the connection between this science and other branches of Physics, some few pages have been set apart to a brief exposition of the Doctrine of Energy.

In arranging the contents of this volume I have deviated considerably from the plan usually adopted, and have been guided by the general principle that the idea of Motion is more elementary than that of Force, and that two Forces, at least, must combine to produce Equilibrium. In accordance with this view the subject of Statics has been made to depend on the laws of Dynamics, and these are preceded by a discussion of some of the simplest principles of Motion. I cannot help thinking that the theory of Equilibrium occupies too prominent a position in many of our Text-books, and that the student obtains, in the problems of Statics, a very inadequate idea of Force and of its modes of expression. In the present arrangement I have followed that order which appears to me to be most logical, and which experience in teaching has shown to be practically advantageous.

The book contains that amount of matter which a pupil may be expected to acquire in a first year's course of instruction in the subject. It is divided into a number of Sections, which may serve as separate lessons, and should be studied in the order in which they occur. All important propositions are illustrated by numerical examples, worked out in the text, and the lessons are furnished with exercises progressively arranged. To render the work more especially serviceable to candidates for the Matriculation and Preliminary Scientific Examinations of the University of London, all questions set during the last few years have been classified, and appended to the several chapters to which they correspond. Answers to the Exercises and Examination Questions are given at the end of the volume.

In writing these lessons, I have had in view the want, which is very generally felt, of a School Text-book, sufficiently elementary to be placed in the hands of a beginner, and yet affording a trustworthy basis for the subsequent work of the student. To what extent I have succeeded in meeting this want, I must leave others to determine.

I take this opportunity to express my thanks to my friend Mr. BENJAMIN KISCH, Barrister-at-Law, for the assistance he has afforded me in revising the proofs, and for many valuable suggestions, whilst the work was passing through the press.

P. M.

January 1875.

PREFACE TO THE SECOND EDITION.

THE DEMAND for a Second Edition within a few months after the appearance of the First, has enabled me to make a few corrections in the Answers to the Examples, and some slight additions to, and alterations in the text, which will, I trust, increase the usefulness of this little work.

P. M.

November 1875.

CONTENTS.

INTRODUCTION.

§ 1. Motion—§ 2. Rest—§ 3. Motion in Molecules—§ 4. Varieties of Motion—§ 5. Physics defined—§ 6. Mechanics defined . . . . Page 1

KINEMATICS—MOTION.

CHAPTER I.

MEASUREMENT OF MOTION.

LESSON

I. UNIFORM AND ACCELERATED—SPACE DESCRIBED.

§ 7. Velocity—§ 8. Uniform Velocity—§ 9. Variable Velocity—§ 10. Examples—§ 11. Space described—§ 12. Examples—§ 13. Space described in a particular Second—§ 14. $V^2 = 2fs$—§ 15. Relation between Final and Initial Velocity and Space described—Exercises 10

II. COMPOSITION OF VELOCITIES.

§ 16. Resultant Velocity—§ 17. Formulæ of Motion—§ 18. Composition of Velocities not in the same Straight Line—§ 19. Parallelogram of Velocities—§ 20. Polygon of Velocities—§ 21. Composition of

LESSON

Uniform and Uniformly Increasing Velocity—§ 22. Law of Composition of Velocities—§ 23. Resolution of Velocity—§ 24. Examples—Exercises Page 20

III. GEOMETRICAL REPRESENTATION OF MOTION.

§ 25. Representation of Time and Velocity—§ 26. Of Space described—§ 27. Of Uniform Motion—§ 28. Of Uniformly Accelerated Motion—Examination 30

CHAPTER II.

FALLING BODIES.

IV. BODIES FALLING FREELY.

§ 29. Application of General Principles of Motion—§ 30. Time of Rising—§ 31. Whole Time of Flight—§ 32. The height to which a body rises—§ 33. Examples 40

V. MOTION ON AN INCLINED PLANE.

§ 34. Acceleration up or down an Inclined Plane—§ 35. Determination of Motion—§ 36. Examples—§ 37. Time of falling down a Chord of Vertical Circle—§ 38. Line of quickest descent from a point to a straight line—Exercises—Examination . 46

DYNAMICS—FORCE.

CHAPTER III.

MEASUREMENT OF FORCE.

VI. MASS—MOMENTUM—UNIT OF FORCE—WEIGHT.

§ 39. Matter, Mass—§ 40. Momentum—§ 41

LESSON

Density—§§ 42-3. Force—§ 44. Unit of Force—§ 45. Gravitation—§ 46. Weight and Mass—§ 47. Centre of Gravity—§ 48. Pressure . Page 58

VII. DYNAMICAL FORMULÆ—ATWOOD'S MACHINE—PROBLEMS.

§ 49. Fundamental Equation of Dynamics—§ 50. Relation between P, W, and f—§ 51. Atwood's Machine—§ 52. Problems—Exercises . 66

VIII. IMPULSIVE FORCES.

§ 53. Measure of an Impulsive Force—§ 54 Examples—Exercises—Examination . . 80

CHAPTER IV.

NEWTON'S LAWS OF MOTION.

IX. THE FIRST AND SECOND LAWS.

§ 55. Laws of Motion—§ 56. Law I.—§ 57. Law II. 90

X. THE THIRD LAW OF MOTION.

§ 58. Statement of the Law—§ 59. Statical Reaction—§ 60. Tension—§ 61. Impact—§ 62. Examples—§ 63. Recoil of a Gun—Examination . 94

CHAPTER V.

ENERGY.

XI. WORK—FRICTION.

§ 64. Definition of Work—§ 65. Unit of Work—§ 66. Measure of Work done—§ 67. Friction—§§ 68-9. Measures of Friction—§ 70. Laws of—§ 71. Co-efficient of—§ 72. Examples . . 102

LESSON
XII. VARIETIES OF ENERGY—CONSERVATION OF ENERGY.

§ 73. Definition of Energy—§ 74. Kinetic and Potential Energy—§§ 75–6. Conversion of Heat into Mechanical Energy, and of Mechanical Energy into Heat—§ 76 (*a*). Dissipation of Energy—§ 77. Relations between Heat and Work—§ 78. Law of Conservation of Energy—§ 79. Totality of Physical Energy—§ 80. Transmutation of Energy—§ 81. Indestructibility of Energy—§ 81. (*a*). Relation between Force, Momentum, and Energy—Exercises—Examination . . . . Page 110

CHAPTER VI.

MACHINES.

XIII. GENERAL REMARKS—APPLICATION OF LAW OF ENERGY.

§ 82. Definition of a Machine—§ 83. Relation between Power, Weight, and Friction—§ 84. Mechanical Advantage and Disadvantage—§ 85. Simple Mechanical Powers 127

XIV. LEVER—WHEEL-AND-AXLE.

§ 86. Principle of the Lever—§ 87. Three kinds of Levers—§ 88. Examples of Levers—§ 89. Weight of Lever Considered—§ 90. Examples—Exercises—§ 91. Wheel-and-Axle; its Mechanical Advantage—§ 92. Windlass—Capstan—§ 93. Toothed Wheels—Exercises 131

XV. THE PULLEY.

§ 94. Pulley described—§ 95. Fixed Pulley—§ 96. Single Moveable Pulley—§ 97. Combinations—First System—§ 98. Weights of Pulleys Considered—§§ 99–100. Second and Third Systems—Exercises 143

XVI. THE INCLINED PLANE—WEDGE—SCREW.

§§ 101–2. Ratio of W to P, when W moves uniformly, and P acts parallel to the Plane—§ 103. Ratio of W to P when P acts parallel to Base of Plane—Case of Weight supported by two Forces —§ 104. Examples—§ 105. The Wedge—§ 106. Wedge Moved by Series of Impulsive Forces— § 107. The Screw described—Ratio of W to P, when P just resists W—Exercises—Examination. . . . . . . Page 154

STATICS—REST.

CHAPTER VII.

THEORY OF EQUILIBRIUM.

LESSON

XVII. General Considerations—Forces in the same Straight Line.

§ 108. Problem of Statics—§ 109. Method of Estimating and Representing Statical Forces— § 110. Forces in the same Straight Line— §§ 111–13. Forces in Equilibrium acting at a Point—Exercises 167

XVIII. Composition of Forces Acting at a Point, but not in the same Straight Line.

§ 114. Parallelogram of Forces—§ 115. Experimental Verification—§ 116. Resultant of several Forces acting at the same Point—§ 117. Formula for Resultant of two Forces—§ 118. Special Cases—§ 119. Examples—Pulley with Cords Inclined—§ 120. Resultant of three Forces acting at a Point, but not in the same Plane— § 121–3. Geometrical Properties of the Resultant 171

LESSON

XIX. GEOMETRICAL CONDITION OF EQUILIBRIUM WHEN TWO OR MORE FORCES ACT AT A POINT.

§ 124. Triangle of Forces—§ 125. Extension of this Proposition—§ 126. Examples—Equilibrium on Inclined Plane—§ 127. Polygon of Forces—Exercises . . . . Page 181

XX. RESOLUTION OF FORCES—ANALYTICAL CONDITIONS OF EQUILIBRIUM WHEN ANY NUMBER OF FORCES ACT AT A POINT.

§§ 128–9. Components of a Force—§ 130. Projection of a Force along a Line—§ 131. Composition of resolved parts of Forces—§ 132. Method of Finding the Resultant by resolving the Forces—§ 133. Examples—§ 134. Conditions of Equilibrium—Exercises 191

XXI. FORCES NOT MEETING AT A POINT—PARALLEL FORCES.

§ 135. Like and Unlike Parallel Forces—§ 136. Resultant of two Parallel Forces—§ 137. Determination of its Position—§ 138. Experimental Verification—§ 139. Examples—§ 140. Resultant of several Parallel Forces—§ 141. Centre of Parallel Forces—§ 142. Condition of Equilibrium—§ 143. Examples—§ 144. Resolution of a Force into Parallel Components—§ 145. Example—§ 146. Couples—Exercises 200

XXII. FORCES PRODUCING ROTATION—MOMENTS.

§ 147. Rotation—§ 148. The two elements on which the Rotatory effect of a Force depends—§ 149. Moment of a Force about a Point—§ 150. Geometrical Representation—§ 151. No Moment about a Point in Line of Action of Force—§ 152. Positive and Negative Moments—§ 153. Moments of Forces about a Point in their Re-

LESSON

sultant—§ 154. Equilibrium of Moments—§ 155. Application to Lever—§ 156. Examples—§ 157. Balances—§ 158. Common Balance—§ 159. False Balance—§ 160. Roman Steelyard—§ 161. Danish Steelyard—§ 162. Examples of other Balances—§§ 163–5. General Properties of Moments—Exercises . . . Page 217

XXIII. GENERAL CONDITIONS OF EQUILIBRIUM OF FORCES IN ONE PLANE—RECAPITULATION.

§ 166. Conditions of Equilibrium for any number of Forces in one Plane—§ 167. Recapitulation—General Results—§ 168. Examples—Examination 233

CHAPTER VIII.

CENTRE OF GRAVITY.

XXIV. PROPERTIES OF CENTRE OF GRAVITY—EQUILIBRIUM OF A BODY ON A HARD SURFACE.

§ 169. Definition of Centre of Gravity—§ 170. Equilibrium of a Body Suspended at a Point—§ 171. Centre of Gravity of Uniform Laminæ—§§ 172–3. Equilibrium of a Body resting on a Hard Surface—§ 174. Three kinds of Equilibrium §§ 175–6. Position of Centre of Gravity in Stable and Unstable Equilibrium—§ 177. Energy of a body in its three States of Equilibrium—§ 178. Limiting Position of a Body on an Inclined Plane 248

XXV. METHODS OF FINDING THE CENTRE OF GRAVITY OF PARTICLES AND BODIES.

§ 179. Centre of Gravity of two Heavy Particles—§§ 180–2. of a number of Heavy Particles—§ 183. Of Symmetrical Bodies—§ 184. Of a Triangle—§ 185. Example—§ 186. Of the

LESSON

Perimeter of a Triangle—§ 187. Of any Rectilinear Figure—§ 188. Examples—Exercises Page 259

XXVI. METHODS OF FINDING THE CENTRE OF GRAVITY OF BODIES JOINED TOGETHER, AND OF PARTS OF BODIES.

§ 189. Centre of Gravity of two Bodies—§§ 190–1. Examples—§ 192. Centre of Gravity of the Frustum of a Body—§ 193. Example—§§ 194–5. Centre of Gravity of a number of Particles—§ 197. Of the Surface of a Cone—§ 199. Of a Pyramid—§ 200. Of a Cone—Exercises—Examination 273

APPENDIX.

A. EXAMINATION PAPERS SET AT VARIOUS INSTITUTIONS 289

B. ANSWERS 305

ELEMENTARY MECHANICS.

INTRODUCTION.

§ 1. **Motion.**—Our earliest observations must have shown us that some things are moving and that others appear to be at rest. We know what motion means when we watch the rising of the sun, the passage of a bird through the air, the waving of the trees in the wind, or the rushing of the waves to the sea-shore. Every variety of matter seems to be endowed with the faculty of movement. The stone falls to the ground, the flower opens its petals to the sun, and living creatures of every size and shape move in endless ways.

§ 2. **Rest.**—We are equally familiar with bodies in a state of rest. Nothing seems more immovable than the earth on which we stand. The various things we see around us,—the books that lie upon our shelves, the pictures that hang upon our walls, are all apparently at rest, and we expect them to remain so unless they happen to be disturbed by some external cause. A little thought will show us that this state of rest is not as simple as it seems. Let us suppose that we are travelling in a railway-carriage, and that another train is moving in the same direction on adjoining rails. After a time it overtakes us and then the two trains move on side by side with equal speed. In this case all sense of motion will be lost; the train at which we are looking and the carriage in which we sit will equally appear to be at rest. This simple illustration is sufficient to make us see that objects may be moving when we suppose them to be stationary, and that the evidence of our senses cannot wholly be trusted. Now the earth on which we stand is in the position of the second train. It is moving round the sun with a considerable velocity; but as we are moving with it and at the same rate, it appears to us to be at rest.

Let us consider, further, the condition of those bodies, which although *absolutely* moving with the earth are, *relatively* to us, at rest. Take the picture hanging on the wall. The picture is suspended by cords which hang over a nail. If these cords were to break, or the nail were to give way, the picture, we know very well, would at once fall to the ground. It appears, therefore, that the picture, although at rest, is really *tending* to fall and is only prevented from obeying its natural tendency by the cords and nail that hold it back. What is true of the picture is true of all things that are in any way supported. Each article of furniture in this room would fall through the floor, if the floor were not strong enough to support it. There is a vessel on the table filled with water, and in the side of the vessel is a cork. The water appears to be motionless. Remove the cork and the water immediately begins to flow out. This water then is endowed with a tendency to motion which the sides of the vessel resisted. The air of the room is seldom still. But suppose for a moment that there is no kind of draught. Let a window or fire-place be opened, let the air be freed in some direction from restraint, and it will at once obey its tendency and begin to move. In these examples no reference has been made to the cause or causes that are supposed to produce the several movements indicated. But the causes themselves do not come within the range of our observing faculties. All that observation teaches us is, that bodies tend to move.

§ 3. **Motion in Molecules.**—If we examine matter more minutely we shall find that it consists of molecules or very small portions, and that among these particles movements are constantly taking place which are separately hidden from the eye, but the total result of which can be discerned. We have all noticed, perhaps, that on a line of rails a certain space is left between the separate pieces at the points where they are joined together. This space is left, because it is found that the length of the rails increases in hot weather and decreases in cold weather; and if the line of rails consisted of one continuous bar of iron fixed at its two extremities, it would become bent and twisted in order to find room for its expansion. If we put a liquid into a glass vessel and place the vessel over the flame of a spirit lamp, we shall very soon observe that the liquid is rising in the vessel, and after a time it will begin to boil, and its particles will be violently agitated. We know how seldom perfect stillness prevails in the atmosphere. The wind is always blowing somewhere. Now the motion of the air is caused directly or indirectly by the sun's heat, and the sun's heat is ever varying in intensity. These examples serve to show that heat is accompanied by motion: but this motion takes place among the particles themselves of which the body consists. The body, as a whole, does not move from place to place; but with every variation in its temperature there is a corresponding movement among the particles which compose it.

We can take another illustration. Most persons know what happens if a stick of sealing-wax be rubbed on flannel and then held over some scraps of paper. The pieces of paper will at once begin to move towards the wax, and may be made to stand on end under its influence. They are electrified, and in that condition they tend to move. Now, we cannot here say to what extent the particles of every body are thus influenced. There may exist opposing tendencies to motion, and in that case no visible effect will be produced. But whenever electricity is developed the particles of the body are agitated, and motion, or the tendency to motion, results.

We have hitherto considered inanimate matter. Let us now see what happens in the animal and vegetable world. A plant or animal may be said to differ from a piece of lifeless matter by its *growth and decay*. Now growth implies a continuous change in the particles of which a body consists. A living organism cannot preserve its old particles and at the same time acquire new. It increases by the decay of old and the substitution of new matter. In this respect animate bodies increase much in the same way as a merchant's capital. A capitalist cannot grow rich by hoarding: on the contrary, he must become daily poorer, for how parsimonious soever he may be, he must consume a portion of what he possesses to support life. It is only by spending money, by buying and selling, by constantly exchanging capital and allowing it to be used by labourers as food, that capital can increase. The same is true of living tissue. In growing it undergoes continuous decay, and the decay is continuously repaired. When the process of reparation does not proceed as rapidly as that of decay, the plant begins to fade and the animal to die. In this continuous decay and reproduction we have a further example of motion among the molecules of bodies.

We thus see that bodies themselves and their molecules are constantly in motion or tending to move: that absolute rest nowhere exists; and that what we call rest, which is really rest relatively to us, can be analysed into counteracted tendencies to motion. As motion is thus universally present, we are sensible of what it is, without being able to define it. It does not admit of explanation; for there is no condition, in which matter exists, that is simpler or more elementary.

§ 4. **Varieties of Motion.**—There are different kinds of motion. Let us see what they are. If a body moves from one place to another place it is said to undergo *translation.* It may move in a straight line or in a curve. The run of a billiard-ball is an example of motion in a straight line. The flight of an arrow and the course of the planets illustrate what is meant by curvilinear motion.

When a body moves about a fixed point or axis, round which the particles describe concentric circles, the body is said to *rotate.* A door rotates on its hinges, a wheel on its axle. It frequently happens that several motions are combined. A body tending to move in two different directions may be found to move in a straight line between them, or to assume a curvilinear motion as the result of these two tendencies. Thus the path of a cricket-ball is a curved line, which is the joint effect of the tendency of the ball to move in the direction in which it was thrown, and of its tendency to fall to the earth. Further the motion of rotation is frequently combined with that of translation. This occurs when a wheel rolls along the ground. The motion of the earth round the sun is the result of two tendencies to motion in different directions, producing a curve called the orbit, and of rotation about a fixed axis passing through the poles. When a body is under the influence of opposing tendencies to motion, which exactly counterbalance one another, it is said to be in equilibrium.

There is another kind of motion, to which the name *undulatory* has been applied. It exists under a variety of forms, but may roughly be described as the movement of a particle to and from a particular point. This displacement of the particle is often called its excursion, and is, in all cases, very small. When a series of particles undergo successively this kind of to-and-fro motion a *wave* is said to be produced. The peculiarity of wave-motion is that although the particles never move beyond the limits of an excursion they appear to undergo translation. If we fix our eyes on a piece of wood floating on a sea-wave, we shall observe that whilst the wave approaches ever nearer to the shore, the piece of wood maintains its position, rising and falling continuously. The apparent motion of translation is the result of the up-and-down movement of each successive particle in its own place. This kind of motion is further illustrated, when the wind sweeps over a field of corn and bends the several ears in succession. In this case it is evident that each stalk of corn cannot move out of its own place, and yet the eye can follow the wave as it passes from one end of the field to the other. In an undulation the particular motion of each particle is well illustrated by the movement of the bob of a pendulum, which vibrates to and from the lowest point in its path.

§ 5. **Physics Defined.**—The science of Physics embraces the consideration of bodies and molecules under every variety of motion, and is subdivided according to the particular effect the several kinds of motion produce upon the senses. Thus the passage of a bird in the air produces a sense of locomotion, the resistance of a heavy body tending to fall to the earth a feeling of pressure. Certain visible motions produce the sensation of heat, others give rise to those of sound, light, electricity, &c.

Physical science is divided into two main branches according as the motion to be considered is the motion of a body as a whole, or of the undulations of the particles of which it consists. It has been suggested to call the one branch of the subject 'Molar Physics,' as treating of *motion in mass*, and the other 'Molecular Physics,' as treating of the *motion of molecules*. The flight of a rifle-bullet, the blow of a cricket-bat, the ascent of a balloon are questions of Molar Physics; whilst problems concerning heat, light or sound belong to the other branch of the subject.

§ 6. **Mechanics.**—The term Mechanics has generally been employed, and is adopted in the present volume, to embrace the science of the motion and equilibrium of bodies. It involves the consideration of matter in its three forms, solid, liquid and gaseous. But the following pages will treat of solid matter only. The subject will be considered under three heads. Under the first will be discussed some elementary principles of motion, apart from the consideration of the quantity or quality of the matter moved. This subject is called Kinematics, or the science of pure motion. Under the second head the matter that is set in motion and the cause, or *force* producing it, will be considered. To this division of the subject the name Dynamics, or the Science of Force, has been given. The third part will embrace certain problems connected with bodies at rest, and these will be discussed under the head of Statics, or the Science of Equilibrium.

KINEMATICS—MOTION.

CHAPTER I.

MEASUREMENT OF MOTION.

I. Uniform—Accelerated Motion—Space described.

§ 7. **Velocity.**—The first question we have to determine is, how motion may be measured and numerically represented. Now we measure all things by their effects, and the visible effect of motion is change of place. When we consider motion with reference to *time* we obtain the idea which is embodied in the word *Velocity*. The introduction of the idea of time distinguishes kinematics from a purely geometrical science. Velocity, or rate of motion, is measured by the amount of change of place, *i.e.* by the space traversed in a given time.

Motion may be uniform or variable. When uniform, equal spaces are described in equal times, and the velocity is constant. In variable motion, the velocity continually changes. In measuring velocity certain units of time and space are adopted. The unit of time is everywhere one second; the unit of length is one foot in England, but is different in different countries.

In England, therefore, the velocity of a moving body is measured by the number of feet traversed in one second; and a body is said to be moving with a unit of velocity when it moves through one foot in one second. The unit of velocity may be briefly indicated as a *foot-second.* When the velocity is variable, it is, at any moment, measured by the space through which the moving body would pass in one second, if it were to continue to move throughout that second with the velocity which it had at that particular moment.

§ 8. **Uniform Velocity.**—If v be the uniform velocity of a moving body, *v equals the number of feet which a body traverses in one second*, and

$2v$ is the number of feet traversed in 2 seconds.
$3v$ " " " 3 "
. . . .
tv " " " t "

If s = space traversed in t seconds, then $s = tv$. This is the fundamental proposition of uniform motion.

Suppose a body to be rotating about a fixed

point, like the sail of a windmill, and that it sweeps out an angle, the measure of which is θ, in one second, then θ is said to be the angular velocity of the body, and $t\,\theta$ will be the angle described in t seconds. If a be the measure of the arc, $t\,a$ will be the space traversed in t seconds.

§ 9. **Variable Velocity.**—The velocity of a body may increase or decrease, uniformly or variably. If it increase or decrease uniformly, the motion is said to be uniformly accelerated or retarded. If the velocity vary, but not uniformly, the motion is said to be un-uniformly accelerated or retarded. As problems connected with variable acceleration are very complicated, requiring for their solution the higher parts of mathematics, we shall consider, in the following pages, uniform acceleration only.

Uniform acceleration or retardation is measured by the increase or decrease of the velocity per second. Thus, suppose a body is found to be moving at the beginning of three successive seconds, with the velocity of 10 ft., 15 ft., and 20 ft. respectively, the body is said to be moving with a uniform acceleration of 5 ft. per second. So too, if a body started with a velocity of 60 ft. per second, and at the end of the first second was moving with a velocity of 50 ft. only, and at the end of the next second with a velocity of 40 ft., the body would be said to be moving with a minus acceleration or retardation of 10 ft. It is clear that all propositions with respect to uniformly increasing velocity are equally true with respect to uniformly decreasing velocity.

Let f represent the acceleration of a moving body, then

The velocity	gained in	1 second	is	f
"	"	2 seconds	is	$2f$
"	"	3	"	$3f$
. . .	. . .	. . .	. . .	. . .
"	"	t	"	tf

If we call v the velocity gained or lost in t seconds when a body is moving with a uniform acceleration of f feet per second, then $v = tf$.

§ 10. **Examples.**—(1) A body starting from rest has been moving for 5 minutes, and has acquired a velocity of 30 miles an hour; what is the acceleration of the body in feet per second?

$$\text{Here, the vel.} = \frac{30 \times 1760 \times 3}{60 \times 60} \text{ ft. per second}$$
$$= 44.$$

$$\text{and } v = tf \therefore 44 = 5 \times 60 \times f \therefore f = \frac{11}{75} \text{ ft. per sec.}$$

(2) If a body move from rest with a uniform acceleration of $\frac{2}{3}$ ft. per sec., how long must it be moving to acquire a velocity of 40 miles an hour?

Here, vel. $= \frac{40 \times 1760 \times 3}{60 \times 60} = \frac{176}{3}$ ft. per sec.

and $f = \frac{2}{3} \therefore \frac{176}{3} = t \times \frac{2}{3} \therefore t = 88$ secs.

(3) What velocity does a body acquire in 3 minutes, if its motion is accelerated at the rate of 32 ft. per sec.?

Here $f = 32$, $t = 3 \times 60$, $\therefore v = 32 \times 3 \times 60 = 5760$.

§ 11. **Space Described.**—To find the space through which a body passes when it moves with a uniform acceleration is a somewhat difficult problem, requiring higher mathematics than we are supposed to have at our command. Later on, we shall show how this problem may be solved by a purely geometrical method, but now we shall content ourselves with explaining how it immediately follows from a simple and almost self-evident proposition.

Suppose a train to pass a certain station at the rate of 20 miles an hour, and to pass another station, one hour afterwards, at the rate of 30 miles an hour, and that the velocity has increased uniformly throughout the interval. Now, it is very evident that those two stations are 25 miles apart; for the *mean* velocity with which the train has moved is 25 miles per hour, and since the velocity has increased uniformly throughout the interval, we may clearly assume that the space described in the first half-hour was as much less than it would have beer, if the velocity had been uniform and 25 miles per hour, as in the second half-hour it was greater. The train will, therefore, have described the same space in the hour, as it would have done if the velocity had been uniform and equal to $\frac{20 + 30}{2}$ *i.e.* 25 miles. Enunciating this principle generally we may say:—*The space described in any given time by a body moving with a uniform acceleration equals the space that it would have described, if it had been moving throughout the given time with a uniform velocity equal to the mean of its initial and terminal velocities.*

We are now able to determine the space described in t seconds, when a body moves with an acceleration of f feet per second.

First. Let the body start from rest. In this case the initial velocity $= o$, and the velocity after t seconds, that is the terminal velocity $= tf$, $\therefore$ mean vel. $= \frac{o + tf}{2} = \frac{tf}{2}$; and if a body move for t seconds with a velocity equal to $\frac{tf}{2}$, the space described is $t \times \frac{tf}{2}$ (§ 8).

$$\therefore s = \frac{t^2 f}{2}.$$

Secondly. Suppose the velocity at starting to

be u, then t seconds afterwards the velocity will be $u + tf$, and the mean velocity is

$$\frac{u + u + tf}{2} = u + \frac{tf}{2}$$

$$\therefore \text{ space described in } t \text{ seconds} = s = tu + \frac{t^2 f}{2}.$$

Generally, if u be the velocity at the beginning of a time t, and v the velocity at the end of that period, the space described in the time $t = t\,\dfrac{u + v}{2}$.

§ 12. **Examples.**—(1) A body commences to move with a velocity of 30 ft. per sec., and its velocity is each second increased by 10 ft. per sec. Find the space described in 5 seconds.

Initial vel. = 30; Final vel. = 30 + 50 = 80

$$\therefore \text{ mean vel.} = \frac{30 + 80}{2} = 55$$

and space described = 5 × 55 = 275 feet.

(2) Find the acceleration, if a body starting with a velocity of 10 ft. per sec. describes 90 ft. in 4 secs.

Let f = acceleration; Initial vel. = 10; Final vel. = $10 + 4f$.

$$\therefore \text{ mean vel.} = \frac{10 + \overline{10 + 4f}}{2} = 10 + 2f.$$

$$\therefore \text{ space described} = 4(10 + 2f) = 40 + 8f = 90 \text{ ft.}$$

$$\therefore f = 6\tfrac{1}{4} \text{ feet per second.}$$

§ 13. **To find the space described in any particular second, when a body moves with a uniform acceleration f.**

If the body start from rest the space described in the first second is $\frac{o + f}{2} = \frac{f}{2}$

Space described in 2nd second is $\frac{f + 2f}{2} = \frac{3f}{2}$

„ 3rd „ $\frac{2f + 3f}{2} = \frac{5f}{2}$

„ 4th „ $\frac{3f + 4f}{2} = \frac{7f}{2}$

Hence, it will be seen, by looking at the numbers 1, 3, 5, 7, that the space described in the tth second equals the tth odd number multiplied by $\frac{f}{2}$

$$\text{or } s = (2t - 1)\frac{f}{2}$$

So also, the space described in the first t seconds of the motion equals the sum of the first t odd numbers multiplied by $\frac{f}{2}$; or $s = \frac{t^2 f}{2}$ as before, (§ 11).

§ 14. From the two formulæ $v = tf$ (§ 9) and $s = \frac{t^2 f}{2}$ (§ 11) we obtain, by eliminating t, a third formula which connects the velocity acquired with the space described. Thus:—

$v = tf$ and $\therefore\ t = \frac{v}{f}$; also $s = \frac{t^2 f}{2} = \frac{v^2}{f^2} \cdot \frac{f}{2} = \frac{v^2}{2f}$

$$\therefore v^2 = 2fs.$$

Examples.—(1) Find the velocity of a body, which starting from rest with an acceleration of 10 feet per sec. has described a space of 20 ft.

$$v^2 = 2fs = 2 \times 10 \times 20 = 400;$$
$$\therefore v = 20.$$

(2) What space must a body traverse to acquire a velocity of 50 ft. per sec. if it move with a uniform acceleration of 5 ft.?

$$50 \times 50 = 2 \times 5 \times s \ \therefore s = 250 \text{ ft.}$$

§ 15. **To find the relation between the velocity and space described, when a body starts with an initial velocity u, and moves with an acceleration f.**

Let h equal the space which the body, starting from rest, would have described under the acceleration f in acquiring the velocity u. Then $u^2 = 2fh$; and if v be the final velocity which the body possesses, after describing the space s, v equals the velocity which the body would have *acquired* if it had started from rest, and passed through $\overline{h + s}$ feet

$$\begin{aligned} \therefore v^2 &= 2f(h + s) \\ &= 2fh + 2fs = u^2 + 2fs \end{aligned}$$
$$\therefore v^2 = u^2 + 2fs.$$

This equation gives the relation between the final velocity, the initial velocity and the space traversed.

Example.—Through what space must a body pass under an acceleration of 5 ft. per sec., so that its velocity may increase from 10 ft. to 20 ft. per sec.?

$$\text{Here } 20^2 = 10^2 + 2 \times 5 \times s;$$

$$\therefore s = \frac{400 - 100}{10} = 30 \text{ ft.}$$

EXERCISES.

1. In what time will a body moving[1] with an acceleration of 25 feet per second acquire a velocity of 1000 feet per second?
2. What space will a body describe in 6 seconds, moving with an acceleration of 160 yards per minute?
3. With what velocity must a body start, if its velocity be retarded 10 feet per second and it come to rest in 12 seconds?
4. In how many seconds will a body describe 1400 feet, moving from rest with acceleration of 7 feet per second?
5. Through what space will a body move in 4 seconds with an acceleration of 32·2 feet per second?
6. A body moving from rest with a uniform acceleration describes 90 feet in the 5th second of its motion, find the acceleration and velocity after 10 seconds.
7. What is the velocity of a particle which moving with an acceleration of 20 feet per second has traversed 1000 feet?

---

[1] In the examples the body is supposed to start from rest, unless otherwise stated.

8. A body is observed to move over 45 feet and 55 feet in 2 successive seconds, find the space it would describe in the 20th second.
9. With what velocity is a body moving after 4 seconds if its acceleration be 10 feet per second?
10. A body moves from rest with an acceleration of 360 yards per minute, and 4 seconds afterwards another body begins to move with an acceleration of ·32 feet per second. When will the latter overtake the former?
11. What velocity must a body have so that, if its velocity be retarded 10 feet per second, it may move over 45 feet?
12. What velocity will be gained by a particle that moves for 5 seconds with an acceleration of 12 feet per second?

II. *Composition of Velocities.*

§ 16. **Resultant Velocity.**—If a body tend to move with several different velocities, the velocity with which it actually moves is called the *resultant*

Fig. 1.

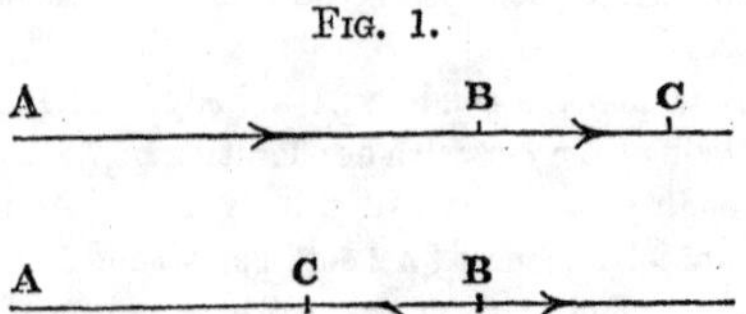

velocity, and those several velocities are called *components.*

If a body tend to move with a velocity u which would take it from A to B in one second, and likewise with a velocity u' which would take it from B to C in the same straight line in one second, then at the end of the second the body will be found at C, as if it had moved with a velocity $u \pm u'$. So too, if the body have several tendencies to uniform motion in the same straight line, the resultant velocity will be the algebraical sum of the component velocities.

Cases of the composition of velocities in the same line occur when a body is moving on something which is itself in motion, as when a boat is descending or ascending a river. Suppose the velocity of the stream to be 3 miles an hour and the vessel to be sailing at the rate of 8 miles an hour in still water, then the actual velocity of the vessel is 5 or 11 miles an hour according as the vessel is sailing up or down stream. When a man paces the deck of a steamer, which is sailing down a river, the actual velocity of the man is the algebraical sum of the velocity of the steamer and of the stream, and of the rate at which the man is walking.

§ 17. If a body tend to move with a uniform velocity and a uniform acceleration[1] in the same straight line, or if it be moving with a certain acceleration along a line, which is, itself, moving uniformly, the resultant velocity at the end of any given time will be the sum or difference of the uniform velocity and the velocity acquired during that time. If u be

[1] The word *acceleration* is used, here and elsewhere, to express a *uniformly increasing velocity.*

the uniform velocity, f the acceleration, and t the time $v = u \pm t f$; and if s_1 be the space the body would describe in t seconds, if moving with the uniform velocity u_1, and s_2 the space it would describe, if moving with the acceleration f_1, then if s be the space actually described, $s = s_1 \pm s_2$

$$= t\,u \pm \frac{t^2 f}{2} \quad (\S\S\ 8,\ 11)$$

If there be several uniform velocities and several accelerations,

$$s = t\,(u_1 + u_2 + \ldots) \pm \frac{t^2\,(f_1 + f_2 + \ldots)}{2}$$

An instance of the composition of a uniform velocity and two different accelerations occurs when a body is projected up or down a rough incline.

The three formulæ, already determined,

$$v = u \pm t\,f$$

$$s = t\,u \pm \frac{t^2 f}{2}$$

$$v^2 = u^2 \pm 2\,f\,s \quad (\S\ 15)$$

are sufficient for the solution of all problems which are concerned with rectilinear motion resulting from the composition of a uniform velocity and uniform acceleration in the same straight line.

§ 18. **Composition of Velocities not in the same straight line.**—Suppose a body tend to move with a uniform velocity u which would take it from A to B in one second, and with a uniform velocity u' which would take it from A to D in one second, then at the end of the second the body will be found at C, where $B\ C$ is equal and parallel to $A\ D$. Moreover, the body will have moved along $A\ C$, and $A\ C$ represents the resultant velocity. That $A\ C$ will be the path of the body may be seen by supposing the body to be moving along $A\ X$ whilst the line $A\ X$ moves parallel to itself with its extremity in $A\ Y$. Then if $A\ B$ be divided into any number of equal parts, say four, and $A\ D$ into

Fig. 2.

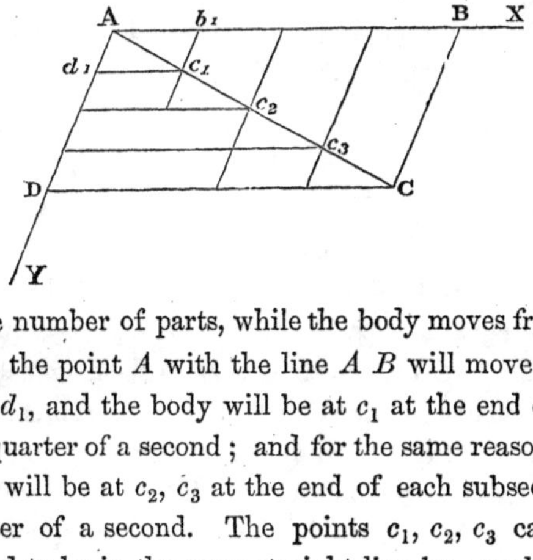

a like number of parts, while the body moves from A to b_1, the point A with the line $A\ B$ will move from A to d_1, and the body will be at c_1 at the end of the first quarter of a second; and for the same reason the body will be at c_2, c_3 at the end of each subsequent quarter of a second. The points c_1, c_2, c_3 can be proved to be in the same straight line by equality of triangles, and since $A\ c_1$, $c_1\ c_2$, $c_2\ c_3$ are equal, the motion along $A\ C$ is uniform.

§ 19. **Parallelogram of Velocities.**—The foregoing proposition is known as the *parallelogram of velocities*, and may be enunciated thus:—*If a body tend to move with two uniform velocities represented by the two sides of a parallelogram, drawn through a fixed point, then the resultant velocity will be represented by the diagonal of this parallelogram that passes through the same point.*

Since the velocities $A\,B$ and $A\,D$ are equivalent to $A\,C$, it is clear that if a body tend to move with three velocities represented by $A\,B$, $A\,D$ and $C\,A$, the body will remain at rest; and since $B\,C$ is equal to $A\,D$, the three velocities that neutralize one another can be represented by $A\,B$, $B\,C$ and $C\,A$,—the three sides of a triangle taken in order.

§ 20. It follows from the foregoing that if a body tend to move simultaneously with several velocities, which would take it (fig. 3) from O to A, from O to B, from O to C, from O to D in one second, and if $A\,B'$ be drawn equal and parallel to $O\,B$, $B'\,C'$ equal and parallel to $O\,C$ and $C'\,D'$ equal and parallel to $O\,D$, then since $O\,B'$ is the diagonal of the parallelogram formed by $O\,A$, $O\,B$, it represents the resultant of these two velocities, and $O\,C'$ represents, for the same reason, the resultant of the velocities $O\,B'$ and $O\,C$, *i.e.* of $O\,A$, $O\,B$, and $O\,C$; and similarly $O\,D'$ represents the final resultant of the several velocities. We see, therefore, that if a body have these several tendencies to motion, it will be found at the end of a second or of any given time at the same point D', as if it had moved first from O to A, then from A to B', thence from B' to C', and finally from C' to D', *i.e.* along the sides of a polygon which respectively represent the velocities. And if the point D' had coincided with O, or the

Fig. 3.

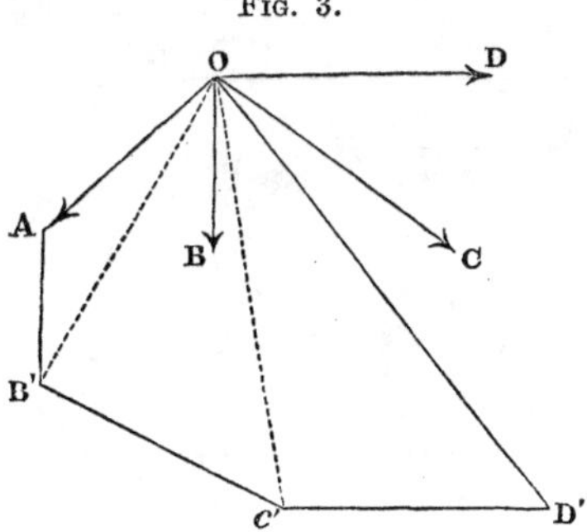

body had had an additional velocity represented in magnitude and direction by $D' O$, the body at the end of the second, or of any less period of time, would have been at O; in other words it would have remained at rest.

If, therefore, the several velocities, with which the body tends to move, can be represented in magnitude and direction by the sides of a closed polygon taken in order, the body will be at rest; but if the velocities are represented by the sides of an open polygon, the body will move, and the resultant velocity will be represented by the straight line that closes the polygon.

§ 21. **Composition of uniform Velocity and Acceleration.**—Suppose a body tend to move with a uniform velocity which would take it from A to B in one second, and likewise with an acceleration that would take it from A to D in one second; then, at

Fig. 4.

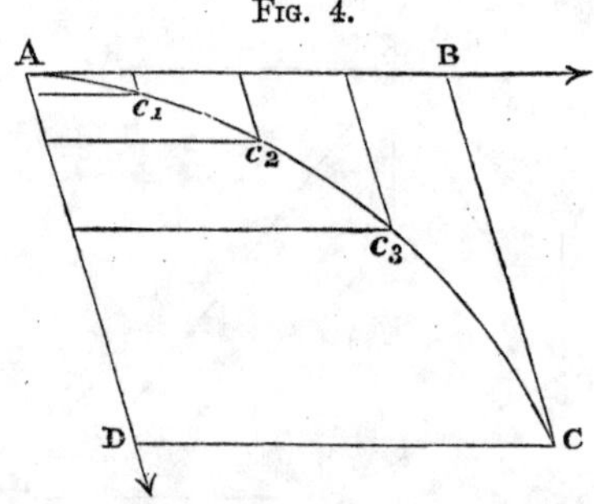

the end of the second the body will be found at C where $B\,C$ is equal and parallel to $A\,D$, just as if it had moved from A to B and from B to C in the second; but the body will *not* have moved along the diagonal $A\,C$. For, since the velocity along $A\,D$ is not uniform, the spaces described in equal intervals of time will not be equal along $A\,D$, whilst they are equal along $A\,B$, and therefore the points c_1, c_2, c_3 will not lie in a straight line. In this case, therefore, the path is a curve, and the nature of the curve depends on the magnitude of the acceleration. The path of a shot projected at a certain angle to the horizon, is a curve resulting from the composition of a uniform velocity in one direction and an acceleration in a different direction.

So, also, if a body tend to move with two different accelerations in different directions, the diagonal of the parallelogram will represent the resultant acceleration, although the path of the body may be along some other line.

§ 22. All these results may be summed up in one general law: *When a body tends to move with several different velocities in different directions, the body will be, at the end of any given time, at the same point, as if it had moved with each velocity separately.* This is the fundamental law of the composition of motions, and it shows that all problems which involve simultaneous tendencies to motion may be treated as if those tendencies were successive.

§ 23. **Resolution of Velocity.**—As the diagonal of the parallelogram, the sides of which represent the component velocities, was found to represent the resultant velocity, so any velocity represented by a certain straight line may be resolved into component velocities represented by the sides of the parallelogram of which that line is the diagonal. Suppose a body urged by a velocity that would take it from O to A in one second, then if $O\,C\,A\,B$ be any parallelogram described on OA as diagonal, the body would equally be at A at the end of one second, if urged by two velocities, which would separately take it from O to C and O to B in one second, and therefore OB, OC represent the components of this velocity.

If OB and OC be at right angles to each other then $OA^2 = OB^2 + OC^2$ and if X be the com-

Fig. 5.

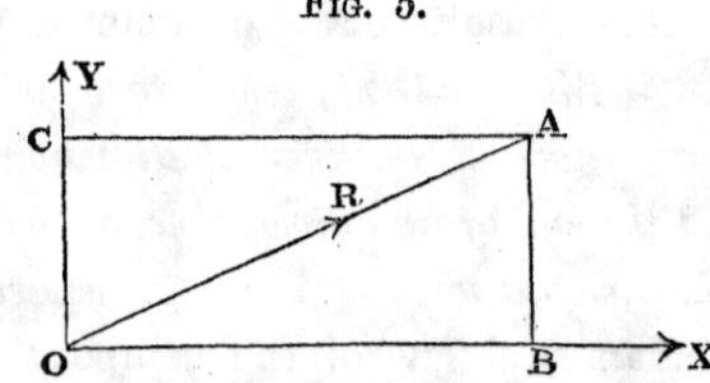

ponent along OB and Y the component along OC, and if R be the original velocity along OA, we have

$$X : R :: OB : OA \text{ or } X = \frac{OB}{OA} R$$

$$Y : R :: OC : OA \text{ or } Y = \frac{OC}{OA} R$$

§ 24. **Examples.**—(1) A body tends to move with velocities of 30 feet and 40 feet per sec. along two straight lines at right angles to each other; find the resultant velocity.

Let $V =$ resultant velocity,
then $V^2 = 30^2 + 40^2 = 2500 \therefore V = 50$ feet per sec.

(2) A body is moving in a certain direction at the rate V feet per second, find the components of its velocity along lines inclined to its direction at angles of 30°, 45°, 60° respectively.

If the angle $A\,O\,B$ is 45° it follows that $O\,B = B\,A$ and $O\,A^2 = 2\,O\,B^2$

$\therefore O\,B = \frac{O\,A}{\sqrt{2}}$, and if V be the velocity along $O\,A$, its component along $O\,B$ is $\frac{V}{\sqrt{2}}$.

Fig. 6.

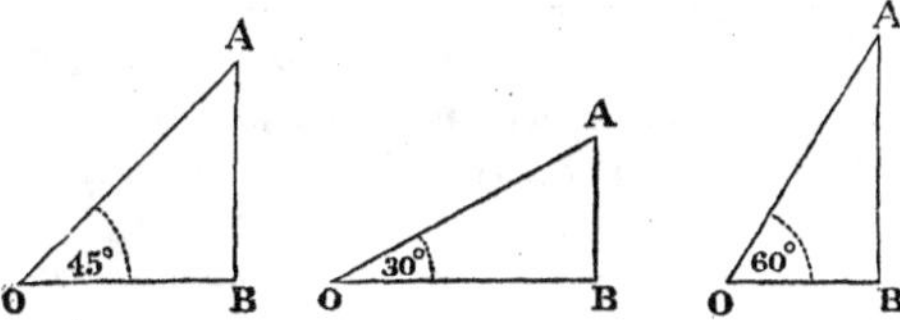

If the angle $A\,O\,B$ is 30°, it follows that $O\,A = 2\,A\,B$,

$\therefore O\,A^2 = O\,B^2 + \frac{O\,A^2}{4}$, or $O\,B = O\,A\frac{\sqrt{3}}{2}$

and the component of the velocity is $\frac{V\sqrt{3}}{2}$.

If the angle $A\,O\,B$ is 60°, $O\,B = \frac{A\,O}{2}$ and the component required is $\frac{V}{2}$.

As these results frequently occur, they should be very carefully remembered.

EXERCISES.

1. A body is simultaneously urged to move with velocities of 50 feet, 21 feet, and 25 feet respectively; can the body remain at rest?
2. A body whilst moving vertically downwards with a uniform velocity of 10 feet per second is urged horizontally with an acceleration of 5 feet per second; find its distance from starting point after 2 seconds.
3. A body tends to move in a certain direction with an acceleration of 32 feet per second, but is constrained to move in a direction inclined at an angle of 45° to the original direction; find the component of its acceleration in the latter direction.
4. A body moving with a uniform velocity of 30 miles an hour has its velocity accelerated 10 feet per second in the same direction; find the space traversed in a quarter of a minute.
5. A body is moving at the rate of 40 miles an hour when its velocity is retarded at the rate of 6 inches per second; when and where will it stop?
6. A body tends to move with equal velocities of 10 feet per second in two directions inclined at 120° to each other; find its path and resultant velocity.

III. *Geometrical Representation of Motion.*

§ 25. In this Lesson we shall show how the subjects that have already been considered, viz., time, uniform and variable velocity, and space described may be geometrically represented.

Let OX be a line limited towards O, unlimited towards X, on which units of length correspond to units of time; so that if the points A, B, C be equally distant from O, and OA represent one second, OB would represent two seconds, and so on. Now let the velocity with which a body is moving at any particular time be represented by a vertical

FIG. 7.

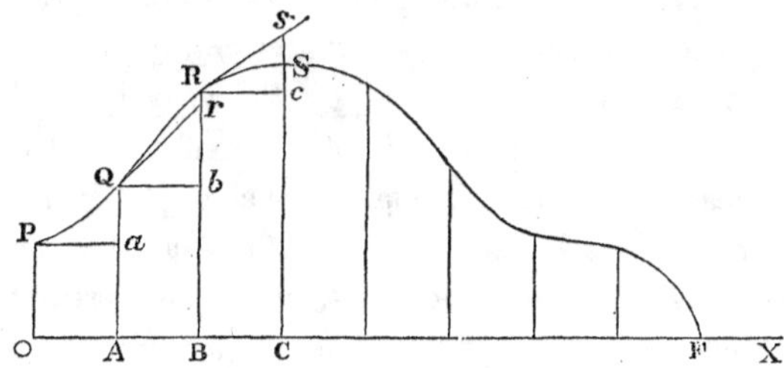

drawn through one of the points O, A, B, C, &c. Thus, if OP represents the velocity at O, and AQ at A, then AQ is greater than OP, in the same proportion as the velocity at A is greater than the velocity at O, and these lines OP, AQ, BR, . . . indicate the number of feet per second with which the body is moving at the points O, A, B. . . . In the same way, if all the corresponding verticals be drawn for periods of time between O, A, B . . . and their extremities be joined, the line $PQRSF$ is called the curve of velocity.

It must not be supposed that the curve of velocity is the same thing as the path of a body.

Motion might take place in a straight line and yet the curve of velocity might be represented by the annexed figure. The curve of velocity is merely a graphic representation of the increase and decrease in the rate of motion at successive intervals of time.

If we draw the lines $P a$, $Q b$ parallel to the line $O X$, $Q a$ and $R b$ will represent the increase of the velocity during the first two seconds; and if this increase were uniform these lines would represent the acceleration. When the acceleration varies, it is measured at any point by the velocity, which would be added in a unit of time, if the velocity increased uniformly throughout such time. If therefore tangents be drawn to the curve at the points Q and R the acceleration at Q would be represented by $r b$, and at R by $s c$. Thus $r b$ measures the rate at which the velocity of the body is increasing per second at the particular moment of time indicated by $O A$, and when the velocity already acquired is $A Q$. In this way we obtain a graphic representation of the velocity and acceleration of the body at any instant of time.

§ 26. We have now to show how the space described in any given time may be graphically represented. Let $O Z$ (fig. 8) represent any interval of time, $A O$ the velocity at O, $Y Z$ the velocity at Z, and $A P Y$ the curve of motion as before. Let $F G$ be a very small interval of time

τ. Let $F P$ be the velocity at the beginning of the time τ, $G Q$ the velocity at the end. Then the space described in the time τ must be greater than the space that would be described if the velocity $P F$ were uniform throughout the interval, and less than the space that would be described with the uniform velocity $Q G$. That is, the true space must lie between $P F \times F G$ and $Q G \times F G$ (since $s = v t$); but $P F \times F G =$ the rectangle $P G$, and $Q G$

FIG. 8.

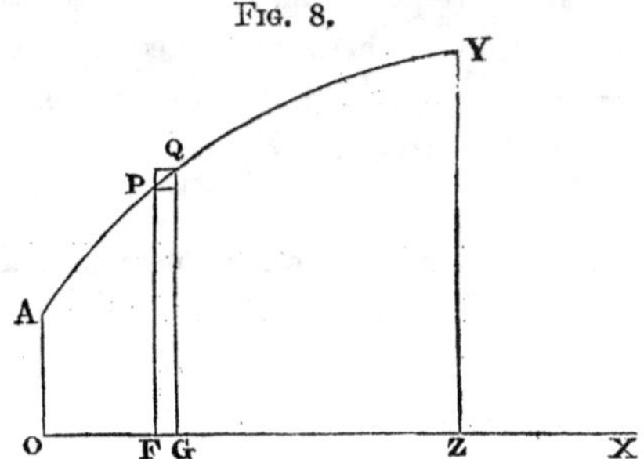

$\times F G =$ the rectangle $Q F$. Therefore the true space described is represented by a figure the magnitude of which lies between the rectangles $P G$ and $Q F$. Now the whole time $O Z$ is made up of the sum of such intervals as $F G$, and therefore the whole space described in the time $O Z$ is somewhere between the number of units of area in the sum of all the rectangles like $P G$, and the sum of all the rectangles like $Q F$. But the sum of each of these sets of rectangles approaches nearer and nearer to the area of the whole figure $O\,A\,Y\,Z$, as $F\,G$ is made smaller and smaller; and can be made to differ from $O\,A\,Y\,Z$ by as small a quantity as ever we please. It thus appears that the space described lies between two quantities; that each of these quantities becomes ultimately equal to $O\,A\,Y\,Z$, as $F\,G$ diminishes without limit; and, therefore, that the space described equals the number of units of area in the figure $O\,A\,Y\,Z$. We have thus proved that the space described in any given time may be represented by the number of units of area contained by the two verticals of velocity, the included line of time, and the portion of the curve intercepted between these two verticals.

The problem of finding the space described in any time resolves itself into that of finding the area of a curve. In all but the simplest cases a knowledge of higher mathematics is necessary.

§ 27. In uniform motion the velocity at different intervals of time remains the same. The curve

FIG. 9.

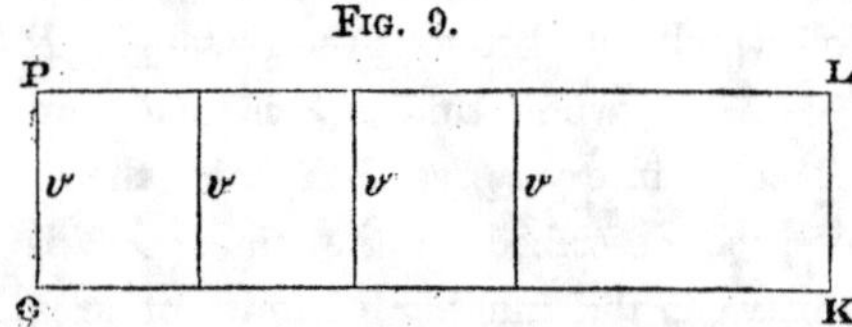

becomes therefore in this case a straight line, parallel to the line of time, and the space described in t seconds equals the area $O\,L = O\,K \times K\,L = t\,.\,v$.

§ 28.—To find the space described in t seconds, when a body moves with a uniform acceleration.

In this case the increments of velocity for successive seconds are constant.

First. Let the body start from rest. Then it OX be the line of time, and OA, AB . . represent

Fig. 10.

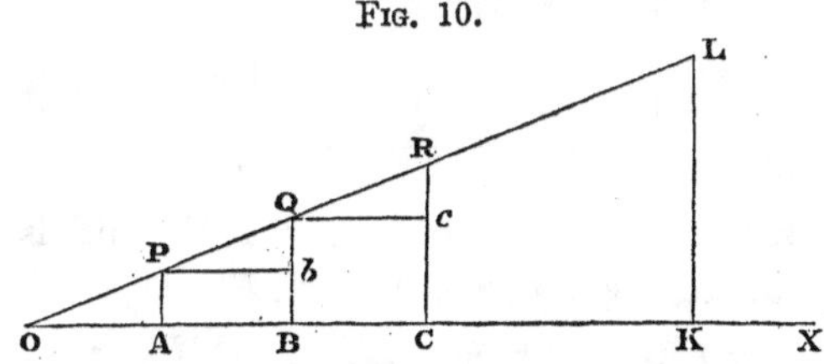

seconds, and if PA represent the velocity at A, QB at B and RC at C; and if Pb, Qc be drawn parallel to OX, then $PA = Qb = Rc = f$, the acceleration, and OPQ can be proved to be a straight line.

Let LK represent the velocity after t seconds, then $OK = t$ and $LK = tf$, and the space described in t seconds equals the area of the triangle $OLK = \frac{t \times tf}{2}$

$$\therefore \; s = \frac{t^2 f}{2}.$$

The same reasoning as was employed to establish the general proposition (§ 26) might have been used to prove this independently.

Secondly. Let the body start with a given velocity.

Let u equal this initial velocity, f the acceleration, and t the time. Then $OP = u$ and $Qa = f$, where OA represents one second. Let $OK = t$. Then the

FIG. 11.

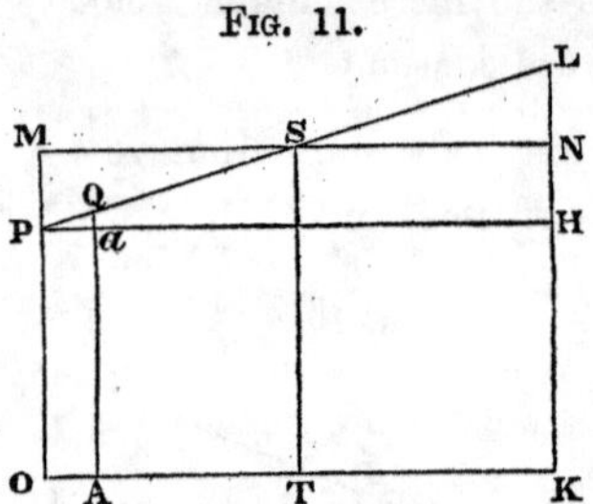

space described is represented by $OPLK$, which is equal to the rectangle OH + triangle PLH

$$= PO \times OK + \tfrac{1}{2} LH \times PH$$
$$= ut + \tfrac{1}{2} tf \cdot t$$
$$\therefore s = tu + \frac{t^2 f}{2}.$$

Let ST bisect OK, and through S draw MSN parallel to OK. Then since the triangle PMS is equal to the triangle SLN, the area

$$OPLK = \text{area } OMNK$$
$$= ST \times OK$$
$$= \frac{OP + LK}{2} \times OK$$
$$= \frac{u + u'}{2} \cdot t.$$

where u' is the terminal velocity.

Or, the space described in any given time, when a

body starts with a certain velocity which is uniformly accelerated, is equal to the space which would have been described, if the body had moved throughout the given time with a uniform velocity equal to the mean of the initial and terminal velocities (§ 11).

EXAMINATION.

1. Distinguish between motion of translation, rotation, and vibration.
2. What is meant by uniform velocity? How is it measured?
3. Two bodies start from *A* to *B* and from *B* to *A*, two points 80 yards apart, at the same time; the one moves uniformly at the rate of 10 ft. per sec., the other at the rate of 12 ft. per sec.; where will they meet?
4. If a particle 10 inches from a given point revolve round it 7 times in 22 secs., find the velocity of the particle.
5. Two men *A* and *B* start at the same moment in the same direction, from two points 1,500 ft. apart; if *A* walk 4 miles an hour and *B* 3½ miles an hour, where will *A* overtake *B*?
6. When is a velocity said to be uniformly accelerated?
7. A body begins to move with a velocity of 100 ft. per sec., and at the end of 7 secs. its velocity is 65 ft. How much is the velocity retarded a second?
8. Show how it is that the space described in any time, when a body moves with a uniform acceleration, is proportional to the square of the time.
9. Enunciate the law of the composition of velocities.
10. A body is simultaneously impressed with three uniform velocities, one of which would cause it to move 10 ft. North in 2 secs., another 12 ft. in one sec. in the same

direction; and a third 21 ft. South in 3 secs. Where will the body be in 5 secs.?

11. Explain the proposition known as the parallelogram of velocities.

12. A body tends to move horizontally with a uniform velocity of 12 ft. per sec., and also vertically downwards with a uniform velocity of 8 ft. per sec.; determine the position after 3 secs.

13. A body begins to move with an acceleration of 8 ft. per. sec., and its velocity is at the same time retarded 2 ft. in 3 secs.; find the space described in 3 secs.

14. Explain why it is dangerous to jump out of a railway carriage in motion.

15. A body is projected horizontally from the top of a cliff with a velocity of 500 ft. per sec.; it reaches the ground in 3 secs.; find its distance from the foot of the cliff.

16. If a person is walking in a straight line, in what direction must he throw a ball upwards, that it may return into his hand?

17. If a ball be thrown out of the window of a railway carriage in motion, in what direction will it seem to fall, and in what direction will it really fall?

18. A body moving uniformly with a velocity of 10 ft. per sec. is suddenly impressed with an acceleration of 32 ft. per sec. in the same direction; what space will be described in the 3rd second of its accelerated motion?

19. A body moves with a velocity of 10 ft. per sec. in a given direction; find the velocity in a direction inclined at an angle of 30° to the original direction.

20. What acceleration along a certain line is equivalent to an acceleration of 20 ft. per sec. in a direction that makes an angle of 45° with that line?

21. A particle moves with a uniformly increasing velocity. Show that the whole space described is proportional to the square of the time from the beginning of the

motion.—(*Matriculation Exam. Univ. Lon.*, Jan. 1871.)

22. A balloon is carried along by a current of air moving from east to west at the rate of 60 miles an hour, having no motion of its own through the air, and a feather is dropped from the balloon. What sort of a path will it appear to describe, as seen by a man in the balloon?—(*Matric.*, June 1874.)

CHAPTER II.

FALLING BODIES.

IV. *Bodies falling freely.*

§ 29. Most of the preceding principles of motion are well illustrated by falling bodies. When a body is allowed to fall freely it is found to acquire a velocity of about 32·2 feet per second every second of its motion, so that it is said to move with an acceleration of 32·2. This acceleration, which for the sake of convenience is represented in books on Mechanics by the letter 'g,' can be shown to vary with the distance of the body from the earth's centre. Thus at the summit of a high mountain g is found to be less than near the surface of the earth; and at the equator, in consequence of the peculiar configuration of the earth, it is less than in the neighbourhood of the poles. Thus the velocity which a body acquires in falling freely for one second varies with the latitude of the place, and likewise with its altitude above the sea-level; but is independent of the size of the body, and of the quantity of matter it contains. Common experience would lead us to suppose that a small ball of lead would fall more quickly than a similar ball of cork, because we are accustomed to see light bodies, such as feathers, fall very slowly to the ground. A little thought, however, will show us that the resistance of the air must have more effect on large and light bodies than on small and heavy bodies, and it may easily be proved by trial that a feather and a ball of lead will fall to the ground in the same time in a vessel from which the air has been removed. Neglecting the resistance of the air, we are able to state that all bodies acquire a velocity of g feet per second in falling to the ground, and that g varies with the distance of the body from the earth's centre, but is the same for all kinds of bodies. As the substance of the body does not need to be taken into consideration, all problems concerning falling bodies may be regarded as cases of accelerated motion in which $f = g$, and may be solved by the application of the formulæ already established:

$$v = u \pm tg$$

$$s = tu \pm \frac{t^2 g}{2}$$

$$v^2 = u^2 \pm 2gs$$

where u is the initial velocity with which a body is projected upwards or downwards.

It is evident that if a body be projected upwards it will lose each second of its motion the same velocity which it would gain if it *fell* freely for one second; for observation teaches us that all bodies tend to move towards the earth with an acceleration of g feet per second. Hence, a body projected upwards may be said to be moving with a negative acceleration or retardation of g feet per second. The consequences which are involved in this proposition we will now proceed to consider separately.

§ 30. **To find the time during which a body rises when projected vertically upwards with a certain velocity.**

Let u be the velocity of projection, then

$$v = u - g\,t$$

gives the velocity of the body after any time t. Now, when the body has reached its highest point, its velocity equals zero, and t is the time occupied in acquiring this velocity. If, therefore, we put $v = o$ in the above equation the corresponding value of t will be the time of rising.

$$\therefore t = \frac{u}{g}.$$

This shows that a body takes the same time to lose a velocity g in rising as to acquire it in falling.

§ 31. **To find the whole time of flight.**

The formula $s = t\,u - \frac{t^2\,g}{2}$ gives the distance of a body from the starting-point after t seconds, when projected vertically upwards with the velocity u. Now it is evident, that when a body has risen to its maximum height and returned to the point of projection, $s = o$. If, therefore, we put $s = o$ in the above equation we get t equal to the whole time of flight;

$$\therefore tu - \frac{t^2 g}{2} = o$$

which gives $t = o$ or $t = \frac{2u}{g}$. The former of these two values shows that $t = o$, before the body starts; the latter that $t = \frac{2u}{g}$, when the body has returned. Hence $\frac{2u}{g}$ is the *whole time* of flight. But $\frac{u}{g}$ has just been proved to be the time of rising, therefore $\frac{u}{g}$ must also be the time of falling; *i.e. the time of rising equals the time of falling.*

If v equal the velocity with which the body passes any point in its path when rising, then, since v at that point may be considered as a velocity of projection, the body will have the same velocity when it returns to that point. In other words, a body passes each point in its path with the same velocity, whether rising or falling. This proposition may be proved directly from the formula

$$v^2 = u^2 - 2gs,$$

where v = velocity with which the body passes a

point the distance of which from the point of projection is s. For, since

$$v^2 = u^2 - 2gs,\ v = \pm\sqrt{(u^2 - 2gs)}.$$

Hence v has two equal values differing in sign only.

§ 32. **To find the height to which a body will rise when projected vertically upwards with a given velocity.**—Take the formula $v^2 = u^2 - 2gs$; then, since $v = o$ at the summit, the corresponding value of s equals the height to which the body will rise,

$$\therefore\quad u^2 = 2gs$$

or

$$s = \frac{u^2}{2g}.$$

Since $u^2 = 2gs$, where s is the height to which a body rises, and u is the velocity of projection, we see that a body would rise through the same space in losing a velocity u, as it would fall through to gain it.

§ 33. **Examples.**—(1) A body projected vertically downwards with a velocity of 20 ft. a sec. from the top of a tower reaches the ground in 2·5 secs.: find the height of the tower.

Here $t = 2\frac{1}{2}$; $u = 20$ and $s = tu + \frac{t^2g}{2}$. Assume $g = 32$.

Then $s = 50 + \frac{16 \times 25}{4} = 150$ ft.

(2) A body is projected vertically upwards with

a velocity of 200 ft. per sec., find the velocity with which it will pass a point 100 ft. above the point of projection.

Here $u = 200$, $s = 100$ and $v^2 = u^2 - 2gs$

$\therefore \quad v^2 = (200)^2 - 64 \times 100$

$= 40{,}000 - 6400 = 33600$

$\therefore \quad v = 40\sqrt{21}$.

(3) A man is rising in a balloon with a uniform velocity of 20 ft. a sec., when he drops a stone which reaches the ground in 4 secs.: find the height of the balloon.

Here $u = 20$, and $t = 4$

and $s = -tu + \dfrac{t^2 g}{2}$

$\therefore s = -80 + 16 \times 16 = 176$

$\therefore$ the height of the balloon was 176 ft.

(4) A shot is fired in a direction inclined at 60° to the horizon, with a velocity of 600 ft. per sec.: find the height to which it rises and its horizontal range.

Here the law of the composition of motion applies. The vertical component of the velocity is $\dfrac{600 \times \sqrt{3}}{2}$ and the horizontal component is 300. If s be the height to which it rises

$$\left(\frac{600 \times \sqrt{3}}{2}\right)^2 = 2gs,$$

$$\therefore s = \frac{300 \times 300 \times 3}{64} = \frac{27 \times 625}{4} = 4{,}218\tfrac{3}{4} \text{ ft.},$$

and if t = time of flight $t = \dfrac{600 \times \sqrt{3}}{32} = \dfrac{75\sqrt{3}}{4}$

∴ the horizontal range is found by seeing how far a body will move horizontally in $\dfrac{75\sqrt{3}}{4}$ seconds with a velocity of 300 ft. Since $s = t\,v$ we have

$$s = \frac{75 \times \sqrt{3}}{4} \times 300 = 75 \times 75 \times \sqrt{3} \text{ ft.}$$

V. *Motion on an Inclined Plane.*

§ 34. Let $A\,B$ be a plane inclined to the horizon at an angle $A\,B\,C$, it is required to find the *accelera-*

Fig. 12.

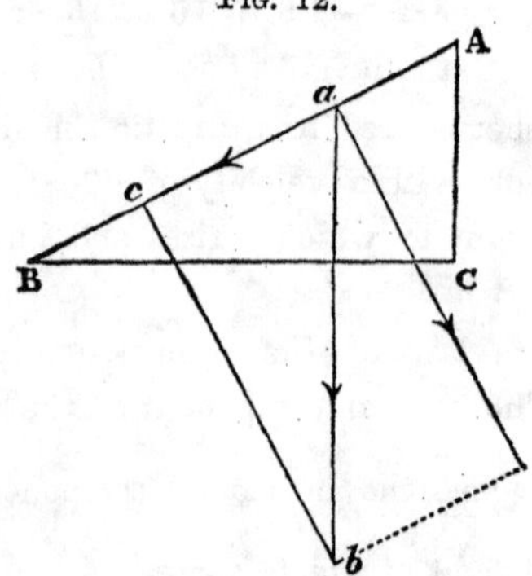

tion with which a heavy body will move, if it rolls down or is projected up the plane.

Let the body be moving down the plane.

Suppose the body to be at a.

Draw $a\,b$ vertically downwards and make it equal to $A\,B$. Let $a\,b$ represent g, the acceleration with which the body would move if free to fall. Then, if $b\,c$ be drawn at right angles to $A\,B$, the triangle $a\,b\,c$ is in every respect equal to the triangle $A\,B\,C$, and $a\,c$ represents the resolved part of the acceleration $a\,b$ in the direction $A\,B$.

Let f be this component of g

Then, $f : g :: a\,c : a\,b$ (§ 23)

$$\text{or,} \quad f = \frac{a\,c}{a\,b} \cdot g = \frac{A\,C}{A\,B} \cdot g.$$

Since $a\,c = A\,C$, and $a\,b = A\,B$.

Let $A\,C$, the height of the plane, equal h, and $A\,B$ the length of the plane, equal l.

$$\text{Then } f = \frac{h}{l} \cdot g$$

If the angle of the plane is $30°$, $f = \frac{g}{2}$

" " $45°$, $f = \frac{g}{\sqrt{2}}$

" " $60°$, $f = \frac{g\sqrt{3}}{2}$

If the body be projected up the plane the retardation due to the body's tendency to fall will also be represented by $a\,c$, and will be equal to $\frac{h}{l} \cdot g$.

§ 35. The motion of the body, if allowed to fall

down the plane, can be ascertained from the equations $v = tf$; $s = \frac{t^2 f}{2}$; $v^2 = 2fs$, by putting $f = \frac{h}{l} g$; and if the body be *projected* up or down the plane, the motion can be determined by substituting this value of f in the equations.

$$v = u \pm tf,\ s = ut \pm \frac{t^2 f}{2};\ v^2 = u^2 \pm 2fs$$

Where u is the velocity of projection.

§ 36. **Examples.**—(1) Find the velocity with which a body must be projected up an inclined plane, the height of which is h feet, and length l feet, to reach the top.

At the summit $v = o$, $\therefore u^2 = 2fs = 2\,\frac{h}{l}.g.l$; $\therefore u = \sqrt{2hg}$, *i.e.* the same velocity as would be needed to project the body from C to A.

(2) Find the time occupied in falling down the whole length of an inclined plane.

$$s = \frac{t^2 f}{2} = \frac{h}{l}.g.\frac{t^2}{2} = l \quad \therefore t^2 = \frac{2l^2}{hg}$$

$$\text{or } t = l\sqrt{\frac{2}{hg}}$$

If the height of the plane remains the same, the time of falling varies directly with the length.

§ 37. **To find the time of falling down any chord of a vertical circle drawn through its highest point.**

Let $A\,C$ be the chord, $A\,B$ the diameter of the

FIG. 13.

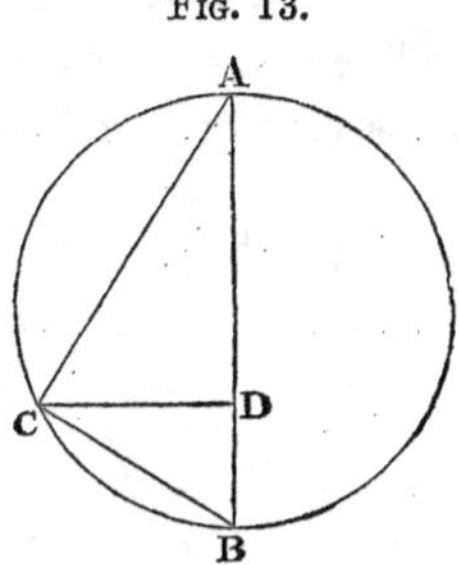

circle. Join $C\,B$ and draw $C\,D$ perpendicular to $A\,B$. Then the acceleration down $A\,C = \frac{A\,D}{A\,C}\,g = \frac{A\,C}{A\,B}\,g$, by similarity of triangles.

$$\therefore \text{ since } s = \frac{t^2 f}{2}, \text{ we have } A\,C = \frac{t^2\,A\,C}{2\,A\,B}\,g$$

$$\therefore \frac{2\,A\,B}{g} = t^2 \;\therefore\; t = \sqrt{\left(\frac{2\,A\,B}{g}\right)}$$

which is constant, and shows that the time of falling down any chord is the same as the time of falling down the diameter.

§ 38. This proposition enables us to find the line of quickest descent from a point to a curve or from one line to another.

The line of quickest descent from a point to a straight line inclined to the horizon may be found thus:—Let P be the point, $A B$ the straight line. Through P draw $P A$, a horizontal line meeting $A B$ in A. Bisect the angle $P A B$ by $A O$. Through P draw $P C$ vertical and from C draw $C Q$ perpendicular to $A B$. $P Q$ is the line of quickest descent. For it is evident that a circle may be described which has

Fig. 14.

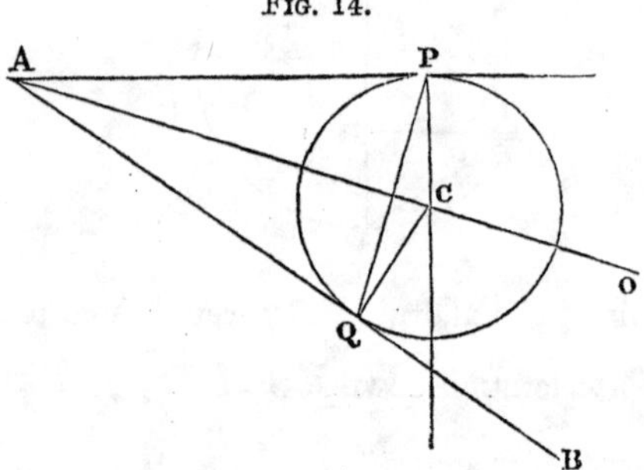

C for its centre and which touches $A B$ and $A P$ in Q and P; and since the time of falling down all chords of this circle from P is the same, $P Q$ must be the line of quickest descent.

The problem of finding the line of quickest descent from a point to a curve is thus found to resolve itself into the purely geometrical problem of drawing a circle, the highest point of which shall be the given point, and which shall touch the given curve.

EXERCISES.

In the following examples g may be supposed to equal 32 ft.

1. Through what distance must a body fall to acquire a velocity of 80 ft. per sec.?
2. A body falls freely for 5 secs.; what is its velocity?
3. A body falls freely for 6 secs.; what space will be described in the last second, and in the whole time?
4. A body is projected upwards with a velocity of 5 g ft.; to what height will it rise?
5. A body is projected upwards with a velocity of 80 ft.; after what time will it return to the hand?
6. A ball is thrown downwards with a velocity of 20 ft. per sec.; find its distance from the point of projection after 3 secs.
7. A ball is thrown downwards with a velocity of 50 ft. per sec.; what is its velocity after 4 secs.?
8. With what velocity must a body be projected vertically upwards that it may rise 40 ft.?
9. A body projected vertically upwards passes a certain point with a velocity of 80 ft. per sec.; how much higher will it ascend?
10. A body projected horizontally from the top of a cliff with a velocity of 40 ft. per sec. strikes the ground after 3 secs.; find the distance of the point of fall from the point of projection.
11. A man descending uniformly the shaft of a mine with a velocity of 100 ft. per min. drops a stone which reaches the bottom in 2 secs; through what distance did it fall?
12. A body starts with a velocity of 90 ft. and loses a third of its velocity per sec.; how far will it move?
13. Two balls are dropped from the top of a tower, one of them 3 secs. before the other; how far will they be be apart 5 secs. after the first was let fall?

14. If a body after having fallen for 3 secs. break a pane of glass and thereby lose one-third of its velocity, find the entire space through which it will have fallen in 4 secs.

15. With what velocity must a body be projected vertically downwards that it may describe 296 ft. in 4 secs.?

16. A body falls freely; find the spaces traversed in the 2nd, 5th, and 7th secs. respectively.

17. A body projected vertically downwards with a velocity of 40 ft. per sec. describes 120 ft. in a certain second; find the space described in the preceding second.

18. Two seconds after a body is let fall another body is projected vertically downwards with a velocity of 100 ft. per sec.; when will it overtake the former?

19. With what velocity must a body be projected vertically upwards to return to the hand after 6 secs.?

20. A ball is dropped from the top of the mast of a ship which is sailing at the rate of 18 miles an hour; if the mast be 64 ft. high, how far will the ship have sailed during the passage of the ball?

21. With what velocity must a ball be projected vertically upwards, to strike the top of a tower 144 ft. high, and how long will it take to reach it?

22. The angle of a plane is 30°; find the velocity with which a body must be projected up it, to reach the top, the length of the plane being 20 ft.

23. Find the velocity of a body that has fallen for 3 secs. down a plane that rises 2 ins. in a length of 5 ins.

24. A body is projected down a plane the inclination of which is 45° with a velocity of 10 ft.; find the space described in $2\frac{1}{2}$ secs.

25. A steam-engine starts on a downward incline of 1 in 200 [1]

---

[1] An incline of 1 in 200 is understood in some books on mechanics to mean 1 foot vertically to a *length* of 200 ft. In other books it signifies 1 foot vertically to 200 ft. *horizontally*.

with a velocity of $7\frac{1}{2}$ miles an hour, neglecting friction; find the space traversed in two minutes,

26. A body projected up an incline of 1 in 100 with a velocity of 15 miles an hour just reaches the summit; find the time occupied.

27. A body is projected with a velocity of 200 ft. per sec. in a direction inclined to the horizon at 45°; find the greatest height it reaches, and the distance of the point where it touches the ground from the point of projection.

EXAMINATION.

1. Find the distance through which a body falls in $1\frac{1}{4}$ sec.
2. Explain, by a graphic representation or otherwise, how it is that a body falling from rest acquires a velocity of 32·2 feet and passes through 16·1 feet only in the first sec.?
3. If a small pebble and a pound of lead be dropped from the top of a tower at the same second, which will reach the ground first?
4. Is there any difference in the velocity which a falling body acquires, when dropped from a certain height near the equator and from the same height near the poles?
5. If a body falls freely, through what space will it pass in the 3rd, 5th, and 7th secs. of its motion?
6. If a body be projected downwards with a velocity of a ft. per. sec., what will be its velocity in x secs.?
7. Prove that a body projected upwards with a certain

Engineers use the expression in the latter sense. We shall adhere to the other signification, as the ratio of the height to the *length* of a plane occurs more frequently than the ratio of the height to the *base.*

velocity takes the same time to lose it, as it would take to gain it, if let fall.

8. A body is projected upwards with a velocity of 100 ft. per sec.; find the whole time of flight.

9. Prove by means of a diagram or otherwise that $s = tu + \frac{t^2 f}{2}$ where u is the initial velocity, and f the acceleration per sec.

10. A ball is projected upwards with a velocity of 150 ft. per sec., and 2 secs. afterwards another ball is projected upwards with a velocity of 200 ft. per sec.; find their distance apart 5 secs. after the first ball was projected.

11. How high will a body rise projected upwards with a velocity of 96·6 ft. per sec.? What will be its velocity $3\frac{1}{2}$ secs. after it was projected? [$g = 32·2$]

12. A body is projected with a velocity of $2g$ ft. per sec. in a direction making an angle of half a right angle with the horizon; show how to find its place at the end of 2 secs.

13. A balloon is rising uniformly with a velocity of 10 ft. per sec., when a man drops from it a stone which reaches the ground in 3 secs.; find the height of the balloon, first, when the stone was dropped; and, secondly, when it reached the ground.

14. A stone falls freely for 3 secs., when it passes through a sheet of glass, in consequence of which it loses half its velocity; find the height of the glass from the ground, if it reaches the earth 2 secs. after breaking the glass.

15. Explain how acceleration, and the space described in any particular second, when a body moves with a constant acceleration, may be graphically represented.

16. A body is projected vertically upwards from the top of an eminence with a velocity of 100 ft. per sec.; find its velocity after 7·5 secs.

17. With what velocity must a body be projected vertically upwards to attain a height of 40,401 ft.?

18. A man is standing on a platform which descends with a uniform acceleration of 5 ft. per sec.; after having descended for 2 secs. he drops a ball; what will be the velocity of the ball after 2 more secs.?

19. A stone projected horizontally with a velocity of 20 ft. per sec. from the top of a tower strikes the ground after 3 secs.; find the distance of the point of fall from the point of projection.

20. Enunciate the principle of motion which is assumed in the solution of the above.

21. A body whilst falling to the ground is attracted horizontally by a force which produces an acceleration of 3 ft. per sec.; find the distance of the body from starting point after 2 secs.

22. Prove that if a body be projected upwards with a certain velocity, the height to which it will rise is equal to the distance through which it would have to fall to acquire that velocity.

23. Show how to find the acceleration with which heavy particles fall down an inclined plane.

24. The angle of a plane is 30°, the length 20 ft.; find the time occupied in falling from the top to the bottom.

25. Show that a body projected vertically upwards will pass all points in its path with the same velocity in rising as in falling.

26. Through what vertical distance must a heavy body fall from rest in order to acquire a velocity of 161 ft. per sec.? If it continue falling for another second after having acquired the above velocity, through what distance will it fall in that time? (Gravity = 32·2.)—*Matriculation, Univ. Lon.*, 1869.

27. A balloon has been ascending vertically at a uniform rate for 4·5 secs., and a stone let fall from it reaches the ground in 7 secs.; find the velocity of the balloon and the height from which the stone is let fall.—*Ib.*, 1869.

28. If a heavy body is thrown vertically up to a given height

and then falls back to the earth, show that, neglecting the resistance of the air, it passes each point of its path with the same velocity when rising and when falling.—*Matriculation, Univ. Lon.*, Jan. 1871.

29. A ball is allowed to fall to the ground from a certain height, and at the same instant another ball is thrown upwards with just sufficient velocity to carry it to the height from which the first one falls; show when and where the two balls will pass each other.—*Ib.*, Jan. 1871.

30. A heavy particle is dropped from a given point, and after it has fallen for one second another particle is dropped from the same point. What is the distance between the two particles when the first has been moving during 5 secs.?—*Ib.*, Jan. 1871.

31. The intensity of gravity at the surface of the planet Jupiter being about 2·6 times as great as it is at the surface of the earth, find approximately the time which a heavy body would occupy in falling from a height of 167 ft. to the surface of Jupiter.—*Ib.*, Jan. 1873.

32. If a body is projected upwards with a velocity of 120 ft. in a second, what is the greatest height to which it will rise, and when will it be moving with a velocity of 40 ft. per sec.?—*Matriculation*, Jan. 1874.

33. Suppose that at the equator a straight hollow tube were thrust vertically down towards the centre of the earth, and that a heavy body were dropped through the centre of such a tube. It would soon strike one side; find which, giving a reason for your reply.—*Ib.*, June 1874.

34. What is meant by saying that the acceleration produced by gravity is represented by the number 32·2?

From a point in a smooth inclined plane a ball is rolled up the plane with a velocity of 16·1 ft. per sec. How far will it roll before it comes to rest, the inclination of the plane to the horizon being 30°? Also, how far will the ball be from the starting-point after 5 secs. from the beginning of motion?—*Matriculation*, June 1874.

35. Prove the formula which gives the distance described in a given time by a body thrown vertically upwards with a given velocity.—*Prelim. Scient.* 1*st M. B.* 1871.

36. A body is projected up an inclined plane with given velocity. Show that the space described in any time is equal to that which would be described in the same time with a uniform velocity equal to half the sum of the velocities at the beginning and end of the time. Hence find the space described on a plane inclined at 30° to the horizon while the velocity changes from 48 to 16 ft. per sec.—*Ib.*, 1872.

37. A heavy body on a level plain has simultaneously communicated to it an upward vertical velocity of 48 ft. per sec., and a horizontal velocity of 25 ft. per sec. Find its greatest height, its range, and its whole time of flight.—*Ib.*, 1874.

DYNAMICS—FORCE.

CHAPTER III.

MEASUREMENT OF FORCE.

VI. *Mass—Momentum—Unit of Force—Weight.*

We have now to consider motion in its connection with the quantity of matter moved, and with the cause producing it.

§ 39. **Matter.—Mass.**—By matter we understand whatever produces resistance. Matter exists in three forms: in the solid, the liquid, and the gaseous form. The earth, the water, and the air are examples of these three conditions of matter. Each offers resistance to the motion of a body. The same substance is sometimes found to exist in Nature under all these forms, which are convertible the one into the other. Thus we are familiar with water in its solid state as ice, in its ordinary liquid state, and as a gas in the form of steam.

The quantity of matter a body contains is called its *mass*.

Mass is invariably measured by weight, and we shall explain later under what conditions this is correct. When we speak of a pound of lead, the word 'pound' expresses a definite quantity of matter, and we suppose all bodies of the same weight to have the same mass. Commercially, weight always stands for mass; and the merchant who estimates his stock by cwts. and tons understands by those weights nothing more than the measure of the quantity of matter he possesses.

§ 40. **Momentum.**—Hitherto we have treated motion apart from the idea of mass, or the quantity of matter moved; and the rate of motion or velocity was in that case its correct measure. But, it is evident that if two bodies are moving with the same velocity there will be a greater quantity of motion in that which contains the greater quantity of matter, just as there is more heat in 10 gallons of water at 10° C. than in one gallon at the same temperature. When the motion of a definite amount of matter is thus considered it is called *momentum*. Thus, the word momentum is employed to express the *quantity of motion in a moving body*, whilst the word velocity is restricted to its original meaning of *intensity or rate of motion*. The difference between momentum and velocity is analogous to that which exists between the quantity of heat a body contains and its temperature. Everyone knows that there is more heat in a hundred gallons of water at 20° C. than there is in a teaspoonful of boiling water, although the temperature of the latter is much higher than that of the former.

In measuring momentum it is necessary to take some fixed amount of motion as a unit. The *unit of momentum* is defined as *the quantity of motion in a unit of mass moving with a unit of velocity.* The unit of mass in this country is the quantity of matter in a standard pound-avoirdupois, and the unit of velocity has been already defined. The unit of momentum is, therefore, the quantity of motion in a body which contains one pound of matter and is moving at the rate of one foot per second.

If a body contain M units of mass and be moving with a unit of velocity, it will possess M units of momentum, and if it be moving with v units of velocity, it will possess Mv units of momentum. This is what is meant by saying, that the momentum of a body whose mass is M and velocity v equals Mv.

§ 41. **Density.**—The quantity of matter in a body does not depend on the size of the body only, but also on the closeness with which the particles are packed. This difference is defined as a difference of *density*. Thus there is more matter in a cubic inch of lead than in a cubic inch of oak, and this is expressed by saying that the density of lead is greater than the density of oak.

§ 42. **Force.**—We are now in a position to consider what is meant by force. The principal properties of matter, with which we are concerned, are, that it moves and offers resistance to the motion of other bodies. Now, force is the name given to the unknown causes of all the various phenomena which matter exhibits: and as all these phenomena are accompanied by motion or the tendency to motion, we shall understand *by force whatever produces or tends to produce motion or change of motion.* It is evident that of forces, *per se*, we can know absolutely nothing. We can only observe their effects; and of these, the most general is motion. We have already seen that matter and the tendency to motion are always conjoined, and this fact has led some writers to identify force and matter. It is quite certain that matter does not exist apart from force; and we need not now pause to consider whether force can exist apart from matter. We shall find it convenient and desirable to consider force as the cause of motion, and wherever we find motion or change of motion we shall assume the existence of force.

§ 43. **Measurement of Force.**—The intimate connection that exists between moving matter and force enables us to measure force by the amount of motion produced or destroyed in a unit of time. Having already defined what is meant by quantity of motion, we see that *the proper measure of force is the momentum generated or destroyed in one second.*

§ 44. **Unit of Force.**—The unit of force is that force which in a unit of time can produce or destroy a unit of momentum. In England, *the unit of force is that force which acting for one second will give to a pound of matter a velocity of one foot per second.* Now, if a body weighing 1 lb. fall freely for one second, it will acquire a velocity of 32·2 feet per second; *i.e.* the force represented by that weight can give to the matter contained in the body 32·2 units of velocity in one second. It follows, therefore, that a pound-weight can produce in one second 32·2 units of momentum, and consists of 32·2 units of force. A weight equal to $\frac{1}{32{\cdot}2}$ lb. is, consequently, sufficient to produce in one second one unit of momentum, and may be considered as the unit of force. This weight may be roughly taken as a half-ounce avoirdupois, so that a half-ounce acting on a pound of matter for one second will give to it a velocity of one foot per second. This fact will, later on, be experimentally verified. As a unit of force can be measured by a certain weight, all forces can be estimated in the same way, and can be represented by equivalent weights.

§ 45. **Force of Gravitation.**—The forces of nature are very various, comprising gravitation, cohesion, heat, electricity, and others. All tend to produce motion, some acting between bodies widely distant in space, and others between the molecules of bodies in intimate juxtaposition. Of these forces that which is most easily measured is gravitation. This force is universally present, and gives rise to the phenomenon known as weight. The force of gravitation acts between all bodies and through any intervening space. The law of universal gravitation was established by Newton. It asserts that every particle of matter attracts, and is attracted to, every other particle of matter, wherever situated; and that the intensity of the force diminishes, as the square of the distance between the two bodies increases. The force of gravitation explains the motion of the planets and other heavenly bodies, as well as the tendency of all bodies near the earth to fall to the ground. Force, being thus universally present in the form of weight, weight becomes a convenient representative of other forces. We can appreciate the power of any other agent, when we compare it with weight, and find out what weight would be able, in the same amount of time, to produce the same effect. It should, however, be clearly understood that weight is but one of a great number of different forces which it serves to measure. Just as money, which is no more wealth than any other commodity, is taken to represent and measure all other kinds of wealth, on account of its easy divisibility, its constancy of value and other properties, so weight is the standard by which other physical forces are estimated. A sovereign may be regarded as a certain amount of wealth in the form of gold, and also as the measure of an equal amount of wealth in some other form; and in the same way a pound-weight, which represents a definite amount of the force of gravitation, serves also to measure other forces capable of producing the same effect. Weight may, perhaps, not inappropriately, be called *the coinage of force.*

§ 46. **Weight and Mass.**—We have seen that by *weight* is understood a certain amount of the force of gravitation, and by *mass* the quantity of matter a body contains. The word weight, however, has been made to do double duty, sometimes standing for force and sometimes for mass. These two significations of the same word should be carefully distinguished. Since the force of gravitation varies inversely with the square of the distance from the earth's centre, the same amount of matter is heavier at one place than at another. Hence, the weight of a body may change whilst its mass remains the same. Thus the same amount of matter weighed in a spring balance would be found to depress the spring less at the equator than at the poles, and less also at the top of a mountain than at the level of the sea. Whenever weight is taken to represent mass it is assumed that the force of gravitation is constant, and on this supposition only can weight be regarded as a correct measure of the quantity of matter a body contains.

§ 47. **Centre of Gravity.**—In connection with the word weight it is well to define a term of frequent occurrence in mechanical problems, and which we shall have afterwards to consider more carefully in its relation to particular bodies. The word 'centre of gravity' is used to express that one point, at which the whole weight of a body may be supposed to act. Such a point is found to exist with respect to every body; and knowing the position of this point we are able to consider the weight of the body as a force acting, at that point, vertically downwards.

§ 48. **Pressure.**—When a heavy body rests on a hard surface and is prevented from producing motion by the molecular attractions between the particles of the substance on which it rests, it is said to exert a *pressure*. Pressure may be produced by other forces than weight; but in all cases the tendency of the force to produce motion is counteracted.

If we try to move a body along a horizontal plane, we shall have to exert a certain amount of force before motion takes place, and this force or pressure is counteracted by the friction between the body and the surface on which it is about to move. If a weight be placed on a sheet of stretched paper it will exert a certain force and then fall through; but if supported on a sheet of metal its tendency to motion will be counteracted, and it will exert, instead, an equivalent pressure. In the theory of equilibrium or Statics pressure is a convenient measure of weight and of such other forces as are prevented from producing motion.

VII.—*Dynamical Formulæ—Atwood's Machine—Problems.*

§ 49. Suppose a force the magnitude of which we will call P is capable of giving to a mass M a velocity of f feet per second, every second of its action, then the force P will generate in one second Mf units of momentum, and Mf would be the measure of P. Since the unit of force is that force which can produce in one second one unit of momentum, the force P which generates in one second Mf units of momentum must contain Mf units of force, or

$$P = Mf.$$

This is the fundamental Equation of Dynamics.

Now, we have just seen that the mass of a body can be measured by its weight only, and that weight is a force which like other forces is measured by the momentum generated or destroyed in one second. Suppose W to be a weight the mass of which is M, at a place where the acceleration due to gravity is g, then the weight W will, if free to descend, acquire in one second a velocity of g feet, *i.e.* the weight W is a force which will generate in one second $M\,g$ units of momentum, and consequently

$$W = M\,g.$$

If we substitute the value of M given by this equation, in the general equation of force, viz. $P = Mf$, we obtain

$$P = \frac{W}{g} . f$$

or $$P : W :: f : g \text{ or } f = \frac{P}{W} . g.$$

This last equation is most important in the solution of dynamical problems, and expresses the fact, that if a force P act for one second on a certain quantity of matter, the weight of which is W, where the acceleration due to gravity is g, then the velocity generated in every second of P's action is f feet per second, where $f = \frac{P}{W} . g$.

In this equation, P stands for any one of the physical forces represented in weight, and is supposed to be acting constantly for some definite period of time. It may mean a muscular force equivalent to a certain number of lbs.-weight, or the elastic force of steam which produces motion in a variety of ways, or the force of an elastic spring which animates a watch, or a certain amount of heat capable of stretching an iron bar. It may also represent a resistance such as friction, which tends to destroy momentum, and which acts in a direction opposite to that in which motion is about to take place. In all cases, if P be the moving or retarding force, and f the velocity produced or destroyed in one second,

$$P = Mf$$

and $$f = \frac{P}{W} \cdot g,$$

where W is the weight of the mass M.

§ 50. The relation that subsists between the quantities of P, W and f, as given in the above equation, is sometimes expressed by the following proposition, which in some books has been incorrectly called the third law of motion: 'When pressure communicates motion to a mass, the acceleration varies directly as the pressure and inversely as the mass.' By pressure is here understood P, the force causing motion, and by mass W, the weight of the mass moved, This law, which follows from the definitions already given, may be verified in many different ways, but most conveniently by a machine called, after its inventor, Atwood's Machine, by means of which we can show that if W remains constant f varies with P, and if P remains constant f varies inversely with W.

§ 51. **Atwood's Machine.**—This machine consists of an upright pillar graduated in feet and inches, and surmounted by a wheel revolving with as little

FIG. 15.

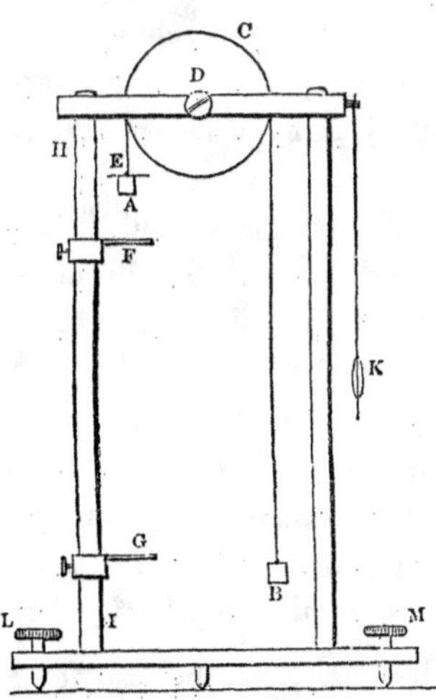

friction as possible. It is also furnished with a pendulum that beats seconds. A fine string passes over the wheel and supports two weights. If these weights be equal, they will either remain at rest or move uniformly with the velocity imparted to them. If a small weight be added to either, that weight becomes the moving force, which together with both weights represents the whole of the mass moved. The additional weight, which serves as a moving force, is generally in the form of a small bar, so that it may remain on the top of a ring with which the upright pillar HI is furnished, and through which the weight A can pass. If now the ring F be placed at such a distance below the starting-point that the weight A reaches it in one second, the velocity then acquired will equal f, the acceleration; and if the bar be left on the top of the ring, the weight will move uniformly with the velocity already acquired. If then a stage G be placed so far below the ring as to stop the weight in one second after the bar has been moved, the distance FG will measure f, the acceleration. Then it will be found that if the weight of the bar be increased whilst the whole mass remains the same, the acceleration will vary directly with the pressure causing motion; and if the weight of the bar remain the same, whilst the equal weights are increased, the acceleration is correspondingly diminished. If, also, the value of f be calculated from the formula $f = \frac{P}{W} g$, the distance FG will be found to accord with it, and in every other respect the experiment will illustrate the results arrived at, by giving values to the symbols in the expression $f = \frac{P}{W} \cdot g$.

The advantage of this machine is that the value

of f can be made as small as we please by taking the weight of the bar sufficiently small compared with the two equal weights, and the value of g may consequently be calculated if f be accurately determined by experiment. Thus, suppose the weight of the bar to be 0·2 ozs., and each of the weights 1-lb., and that the distance FG is found to be 0·2 feet. Then from the formula $f = \frac{P}{W} \cdot g$, since $P =$ 0·2 ozs. and $W = (2 \times 16 + 0{\cdot}2)$ ozs. and $f = 0{\cdot}2$ feet, we have

$$0{\cdot}2 = \frac{0{\cdot}2}{32{\cdot}2} \cdot g$$

or

$$g = 32{\cdot}2 \text{ feet.}$$

§ 52. **Examples.**—(1) Two weights P and Q hang over a smooth wheel, as in Atwood's machine; find the acceleration.

Here the force causing motion is the difference between P and Q, since there would be no motion if P were equal to Q. The moving force is, therefore, $P - Q$. The weight of the mass moved is $P + Q$

$$\therefore f = \frac{P-Q}{P+Q} \cdot g.$$

As soon as the acceleration is determined, all problems connected with motion can be solved by substituting its value for f in the equations

$$v = u \pm tf$$

$$s = tu \pm \frac{t^2 f}{2}$$

$$v^2 = u^2 \pm 2 f s.$$

(2) To find the space described by either weight in 3 seconds, when $P = 5$ ozs. and $Q = 3$ ozs.

Here force producing motion is $5 - 3 = 2$ ozs.; and the weight of the mass moved is $5 + 3 = 8$ ozs.

$$\therefore f = \frac{2}{8} 32 = 8; \text{ also}$$

$$s = \frac{t^2 f}{2} = 9 \times 4 = 36 \text{ feet.}$$

(3) A body weighing 12 ozs. rests on a perfectly smooth horizontal table and is drawn along by a

FIG. 16.

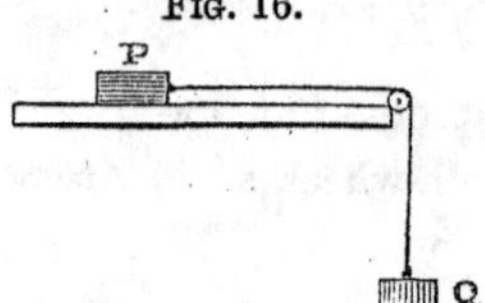

weight of 4 ozs., attached to it by a string that passes over a pulley at the edge; find the velocity after 4 seconds.

Since the weight P is entirely supported by the resistance of the table, the moving force is the weight of 4 ozs. hanging vertically downwards, and the weight of the mass moved $= 4 + 12 = 16$ ozs.

$$\therefore f = \frac{4}{16} 32 = 8; \text{ and } v = tf = 4 \times 8 = 32.$$

(4) Two weights of 9 ozs. and 7 ozs. hang over a

smooth wheel; motion continues for 5 secs., when the string breaks; find the height to which the lighter weight will rise after the breakage.

Here $P = 9 - 7 = 2$ ozs. and $W = 9 + 7 = 16$ ozs.

$$\therefore f = \frac{2}{16} 32 = 4\text{; and } v = tf = 5 \times 4 = 20$$

$\therefore$ each weight has an initial velocity of 20 feet when the string breaks.

$$\therefore \; v^2 = \overline{20}^2 = 2 \times 32 \times s$$

$$\therefore \quad s = \frac{400}{64} = 6\tfrac{1}{4} \text{ feet, } i.e. \text{ the lighter}$$

weight will rise $6\frac{1}{4}$ feet, before it begins to descend.

(5) A steam-engine is moving at the rate of 30 miles an hour when the steam is turned off; supposing the friction to be equivalent to a retarding force of $\frac{1}{400}$ of the weight of the engine, find how long and how far it will move before it stops.

Let W be weight of engine, then if P be the retarding force,

$$P = \frac{W}{400} \text{ and } f = \frac{\frac{W}{400}}{W} g = \frac{g}{400}.$$

The velocity is 30 miles an hour, *i.e.* $\dfrac{1760 \times 3 \times 30}{60 \times 60}$

$= 44$ feet per second. The question is, in how long a time will this velocity be destroyed, if the velocity be retarded $\dfrac{32}{400}$ feet per sec.? Since $v = tf$ we have

$$44 = t\,\frac{32}{400} \text{ or } t = \frac{17600}{32} = 550 \text{ secs.}$$

also $v^2 = 2fs \quad \therefore\ 44 \times 44 = \dfrac{64}{400}s$

$$\text{or } s = \frac{44 \times 44 \times 400}{64} = 12{,}100 \text{ feet.}$$

(6) For how long a time must a force of 3 ozs. act on a mass, the weight of which is 12 ozs., to generate a velocity of 40 feet per sec.?

$$\text{Here } f = \frac{3}{12}\,32 = 8, \text{ and } v = tf$$

$$\therefore\ 40 = 8t \text{ or } t = 5 \text{ secs.}$$

(7) A force of 4 ozs. causes a certain mass to move from rest, through 18 feet in 3 secs.; find the weight of the mass.

$$s = \frac{t^2 f}{2} \quad \therefore\ 18 = \frac{9}{2}\cdot f \quad \therefore\ f = 4$$

and $$f = \frac{P}{W}\cdot g$$

$$\therefore\ 4 = \frac{4}{W}\cdot 32 \quad \therefore\ W = 32 \text{ ozs.}$$

(8) A body weighing 12 ozs. is moving along a rough table, on which the friction is equivalent to a force of 3 ozs., with a velocity of 20 feet per second. After one second, it reaches the edge of the table and falls to the ground in 2 secs.; find the distance of the point of fall from the edge of the table.

Here the retarding force is 3 ozs., and the

weight of the mass retarded is 12 ozs. $\therefore f = \frac{3}{12} 32 = 8$.

If v be the velocity after one second, $v = u - tf = 20 - 8 = 12$.

The body, therefore, leaves the table with a horizontal velocity equal to 12 ft. per sec. It reaches the ground in 2 secs., in which time it will have fallen through $\frac{32 \times 2^2}{2} = 64$ ft. In 2 secs. it will have moved horizontally $2 \times 12 = 24$ ft.

$\therefore$ distance required $= \sqrt{(\overline{64}|^2 + \overline{24}|^2)} = 68$ ft. nearly.

(9) A plane supporting a weight of 12 ozs. is descending with a uniform acceleration of 10 ft. per sec.; find the pressure that the weight exerts on the plane.

When a body rests on a horizontal surface it exerts a pressure equal to its own weight. If the plane moves with an acceleration of 32·2 ft., the weight exerts no pressure whatever. This would be the case if we were to place a weight on a book, and then let the book fall. During the motion, the weight would exert no pressure on the book, since they would both move together.

Let P be the pressure the body exerts on the plane when moving with an acceleration of 10 ft. per second. Then the pressure causing motion is

$(12-P)$ ozs., and the weight of the mass moved is 12 ozs. (neglecting the weight of the plane)

$$\therefore f = 10 = \frac{12 - P}{12} \times 32, \text{ or } P = 8\tfrac{1}{4} \text{ ozs.}$$

(10) A body weighing 50 lbs. is acted on by a constant force which acts for 5 secs. and then ceases to act; the body moves through 60 feet in the next 2 secs. Express the force in absolute units.

The absolute unit of force is g times as great as the unit we have adopted, being that force which will generate in one second a velocity of g feet per second.

Here $W = 50$ lbs. and $f = \frac{P}{W} \cdot g = \frac{P g}{50}$.

Since this force P acts for 5 seconds, the velocity acquired will be $5 f = \frac{5 P g}{50} = \frac{P g}{10}$. The body is then found to move, with this velocity, through 60 feet in 2 seconds. Since $s = t v$ we have

$$60 = 2 \times \frac{P g}{10} \therefore P g = 300$$

the measure of the force in *absolute units.*

(11) A weight Q supported on an inclined plane, which rises h in l, is pulled up the plane by a weight P connected with Q by a thread, which passes over a wheel at the summit of the plane, and hangs vertically downwards. Find the acceleration on the plane.

Let F equal the force with which the body, whose weight is Q, tends to move down the plane, and f its acceleration in that direction, then

$$f = \frac{F}{Q} g \text{ and } \therefore \frac{f}{g} = \frac{F}{Q}$$

But $\quad \frac{f}{g} = \frac{h}{l}$ (§ 34) $\therefore \ F = \frac{h}{l} Q.$

FIG. 17.

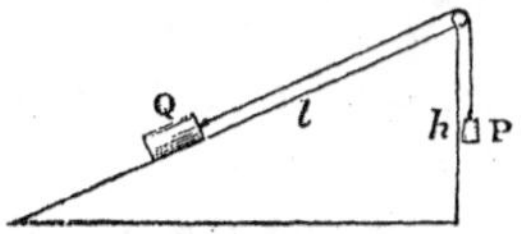

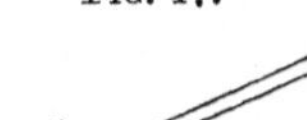

If, therefore, P cause Q to *ascend* the plane, the force causing motion must be $P - F = P - \frac{h}{l} Q$

and the acceleration $= \dfrac{P - \frac{h}{l} Q}{P + Q} \cdot g.$

(12) Two weights P and Q are supported on two inclined planes, the lengths of which are l and l' and the common height h. They are connected by a fine thread as before: find the acceleration.

If F equal the force with which P tends to descend, then $F = \frac{h}{l} P$ as above, and if F' equal the force with which Q tends to descend, then $F' = \frac{h}{l} Q$, and if P is able to draw up Q, the force causing motion is $F - F' = \frac{h}{l} P - \frac{h}{l'} Q$, and the acceleration equals $\frac{h}{l l'} \cdot \frac{P l' - Q l}{P + Q} \cdot g$.

(13) A body is projected up an inclined plane which rises 1 in 10 with a velocity of 40 feet per second. Supposing the effect of friction to be equivalent to a uniformly retarding force equal to $\frac{1}{100}$ the weight of the body, find how far the body will move up the plane.

Here there are two causes tending to destroy the velocity of projection; first, the retardation due to gravity; secondly, that due to friction.

$$\text{Retardation due to gravity} = \frac{h}{l} g = \frac{1}{10} g$$

$$\text{,, ,, ,, friction} = \frac{\frac{W}{100}}{W} \cdot g = \frac{g}{100}$$

$\therefore$ resultant retardation equals $\left(\frac{1}{10} + \frac{1}{100}\right) g$.

Since $v^2 = 2 f s$ we have

$$(40)^2 = 2\left(\frac{1}{10} + \frac{1}{100}\right) g \cdot s$$

$$= \frac{22}{100} \cdot g \cdot s,$$

$$\therefore s = 227\frac{3}{11} \text{ ft.}$$

EXERCISES.

1. Two weights of $4\frac{1}{2}$ ozs. and $3\frac{1}{2}$ ozs. hang over a pulley; find the space moved through from rest in 4 secs.
2. For how long a time must a force[1] of 2 ozs. act on a mass of 4 lbs. that it may give to the mass a velocity of 60 ft. per sec.?
3. What weight hanging vertically downwards will draw a weight of $3\frac{1}{2}$ lbs., across a perfectly smooth table 8 ft. wide in 2 secs.?
4. Through what distance must a force of 3 ozs. act on a mass of 16 ozs. to give to it a velocity of 6 ft. per sec.?
5. Two weights hang over a pulley; the heavier is 12 ozs., and it moves the lighter through 36 ft. in 3 secs.; find the lighter weight.
6. What is the force of friction if a body weighing 20 ozs. projected along a rough horizontal plane with a velocity of 48 ft. per sec., come to rest after 5 secs.?
7. What weight must be added to one of two equal weights of 6 ozs., which hang over a smooth wheel so that they may move through 16 ft. in 5 secs.?
8. Required the force producing motion, if a weight of 25 ozs. be caused to move through 320 ft. in 10 secs.
9. The velocity of a body weighing 12 ozs., increases from 10 ft. per sec. to 20 ft. per sec. whilst the body passes over 15 ft.; what is the moving force?
10. Two weights, one of which is 4 ozs., hang over a pulley; the weight of 4 ozs. ascends with a uniform acceleration of 5 ft. ser sec. Find the other weight.
11. Two weights of $7\frac{3}{4}$ ozs. and $8\frac{1}{4}$ ozs. connected by a thread

---

[1] A *force* of 2 ozs. means a force that produces the same effect as a weight of 2 ozs. acting vertically downwards. A *mass* of 4 lbs. means a mass the weight of which is 4 lbs. where the acceleration due to gravity is g.

hang over a pulley; motion continues for 3 secs., when the string breaks. To what height will the lighter weight ascend, and how far will the heavier weight fall in 4 secs. after the breaking of the string?

12. A steam-engine is moving at the rate of 20 miles an hour when the steam is let off; if the force of friction be equivalent to $\frac{1}{320}$ of the weight of the engine, after what time will it stop?

13. A plane sustaining a weight of 20 ozs. is descending with a uniform acceleration of 10 ft; find the pressure on the plane.

14. A weight of 1 cwt. goes up and down on a lift with a uniform acceleration of 4 ft. per sec.; find the pressure on the lift in each case.

15. A weight of 20 ozs. is moved on a smooth horizontal table by a weight of 4 ozs. connected with it by a string which passes over a pulley at the edge of the table and hangs vertically downwards. After 3 secs. the weight reaches the edge of the table, and the string breaks. At what distance from the top of the table will the weight of 20 ozs. strike the ground supposing it to reach the ground in 1 sec.?

16. If a force of 6 lbs. gives to a certain mass a velocity of 20 ft. per sec. in 4 secs., what velocity will a force of 4 lbs. give to twice the mass in 6 secs.?

17. For how long a time must a force of 3 lbs. act on a mass of 120 lbs. to give to it a velocity of 960 yards per minute?

VIII.—*Impulsive Forces.*

§ 53. The forces we have hitherto considered have been such as, acting for one second, generate a certain amount of momentum, by which they have been numerically expressed. We have now to consider a class of forces which act for only a very short period of time; as, for instance, when a ball is projected from the hand, or struck by a bat, when an arrow is shot from a bow, or a bullet from a gun, when a nail is hammered into a wall, or when piles are driven into the ground. These forces have been called *impulsive forces*, and we have retained the name at the head of this article because it is in such common use; but there is no difference between the *character* of these forces and those which act for a longer time. If we knew the *law* and *duration* of action we could determine all the circumstances in regard to them the same as for other forces. But the *law of action* in these cases is not generally known. When a nail is driven into a wall the hammer and nail are compressed, and the timber yields, but according to unknown laws. We are, however, able to determine certain final results, and these we call *Impulses*. When bodies are free to move the result is a certain momentum; just as the final result of a constant force acting for a finite time is a momentum. If an *impulse* were strictly instantaneous (that is, requiring no time for action) it would cause a body to start from rest and acquire a velocity v without moving the body, since it would have no time in which to move, but no such action ever takes place.

If P be the magnitude of an *impulse*, which causes a mass M to move with a velocity v, then $P = Mv$, because Mv units of momentum are generated by P. Hence $v = \frac{P}{M}$; but $M = \frac{W}{g}$; consequently $v = \frac{P}{M} g$.

This equation differs in form from that previously obtained for constant forces, in so far only, that v is substituted for f. In the former equation f represents the velocity generated in one second, while in the latter v represents the final velocity regardless of the time. Constant forces may be uniform, that is, have a constant or uniform *intensity*, such as we have considered in the preceding pages; or they may constantly vary in intensity, in which case f represents the velocity which it would produce in a second of time if the intensity remained constant during that time; and the expression $M f$ measures the intensity of the force at any instant. For uniform constant forces we have $P=Mf$ (§ 49); multiplying by t we have $Pt=Mft$; but for time t we have $v=f t$ (§ 9); and by substitution we have $P t=M v$; hence $M v$ is a measure of the final *effect* of a *uniform constant* force P, for a time t, as well as that of an *impulse* in which the intensity of the force and the time of action are both unknown. In the former, by knowing the values of P and t, we may find the value of the product $M v$, but in the latter we must know M and v in order to find the value of the sum of the products of $P t$.

§ 54. **Examples.**—(1) If a gun with a certain charge of powder cause a hundred-pound shot to ascend vertically for 3 seconds, through what height will a fifty pound shot ascend with three times as great a charge.

Let P be the impulse of the powder in first case, and P' that in second case.

Then if v be the velocity with which the shot

leaves the gun in first case, and if v' be the velocity with which the shot leaves the gun in second case,

$$v = \frac{P}{100} \cdot g \text{ and } v' = \frac{P'}{50} g$$

but $$P' = 3P \therefore v' = \frac{3P}{50} \cdot g; \therefore \frac{v}{v'} = \frac{1}{6}$$

and, since the shot ascends in the first case for 3 secs. $v = 3g \therefore v' = 18\, g = \sqrt{2gs}$ where s is the height required

$$\therefore s = 5184 \text{ feet nearly.}$$

It will be seen that since $v = \frac{P}{W} \cdot g$, the velocity of projection varies directly with the impelling force and inversely with the weight of the mass projected. And, since the height to which a body rises varies with the square of the velocity of projection, as is seen from the equation $v^2 = 2gs$ or $s = \frac{v^2}{2g}$, the height to which a body can be projected varies directly with the square of the impelling force, and inversely with the square of the mass projected. Remembering this fact, many problems may be solved by the ordinary process of double rule of three.

(2) A shot weighing 2 ozs., and projected with p ozs. of powder, rises to a height of 100 ft.; to what height will a shot rise, that weighs 5 ozs., and is projected with $3p$ ozs. of powder?

A shot weighing 2 ozs. projected with p ozs. of powder rises 100 ft.; therefore a shot weighing 1 oz. projected with p ozs. of powder would rise 100×2^2 ft.; therefore a shot weighing 5 ozs. projected with p ozs. of powder would rise $\frac{100 \times 2^2}{5^2}$ ft.; therefore a shot weighing 5 ozs. projected with 1 oz. of powder would rise $\frac{100 \times 2^2}{5^2 \times p^2}$ ft.; therefore a shot weighing 5 ozs. projected with $3p$ ozs. of powder would rise

$$\frac{100 \times 2^2 \times 3^2 p^2}{5^2 \times p^2} = \frac{100 \times 4 \times 9}{25} = 144 \text{ ft.}$$

EXERCISES.

1. An arrow shot from a bow starts off with a velocity of 120 ft. per sec.; with what velocity will an arrow twice as heavy leave the bow, if sent off with three times the force?
2. If a ball fired from a gun rise to a height of 150 ft., to what height will a ball half again as heavy rise, if fired with twice the charge of powder?
3. Two balls weighing 8 ozs. and 6 ozs. respectively are simultaneously projected upwards, and the former rises to a height of 324 ft. and the latter to 256 ft.; compare the forces of projection.

EXAMINATION.

1. Explain how force is measured.
2. What is the unit of force?
3. On what does the *weight* of a body depend?
4. Two bodies the masses of which were as 3 : 2, were found by a spring balance to weigh in two different places 963 grammes and 628 grammes respectively. Compare the velocities acquired by a body falling for 1 sec. in each of the two places.
5. Explain what is meant by *momentum*, and how it differs from *velocity*.
6. If the unit of length were a yard and the unit of time were a minute, how would the value of g be correspondingly changed?
7. Two weights 15·25 ozs. and 16·75 hang over a pulley; find their velocity after 4 secs.
8. What are the uses of Atwood's machine?
9. What is the law connecting the pressure that communicates motion to a mass, with the acceleration with which the mass moves?
10. Find in what time a force of 5 lbs. will move a weight of 16 lbs. through 45 ft. along a smooth horizontal plane.
11. Find in wnat time a weight of 5 lbs. hanging vertically downwards will move a weight of 11 lbs., with which it is connected by a string passing over a pulley through 45 feet along a smooth horizontal plane.
12. Two bodies whose masses are m and 3 m are moved by forces 3 p and p respectively; compare the spaces they describe in t secs.
13. What force must act on a mass of 48 lbs. to increase its velocity from 30 ft. to 40 ft., whilst it passes over 80 ft.?
14. For how long a time must a force of 1 lb. act on a mass

the weight of which is 40 lbs. to give to it a velocity of 500 ft. per sec.?

15. A weight of 100 lbs. is moving horizontally with a velocity of 20 ft. per sec., and is retarded by friction which is equivalent to a force of 5 lbs. How far will it move?

16. To one end of a string hanging over a pulley is attached a weight of 5 ozs. and to the other end two weights of 3 ozs. and 4 ozs. Motion takes place for 3 secs., when the weight of 4 ozs. is removed; for how long will the weight of 5 ozs. continue to ascend?

17. Two weights P and Q hang over a smooth wheel, connected by a string, and P descends 18 ft. in 3 secs. If, however, 5 ozs. had been added to Q, Q would have descended through the same space in $4\frac{1}{2}$ secs.; find P and Q.

18. How may an impulsive force be measured? If P be the measure of a certain blow, explain clearly the meaning of the equation $f = \frac{P}{W}g$.

19. What is the measure of the blow given to a ball weighing 1 lb. if it start off with a velocity of 224 ft. per sec.?

20. Two balls whose masses are as 2 : 3 are projected vertically upwards with the same force; compare the heights to which they rise.

21. A body of given mass is acted upon by a constant force; find the space described in a given time.—*Matriculation*, Jan. 1870.

22. What is meant by the 'acceleration' due to a force; and upon what does its magnitude depend? If the velocity of a body increases from 12 ft. to 13 ft. per sec. while it moves over a distance of 5 ft., what is the acceleration? Indicate the course of the reasoning upon which your calculation is based.—*Ib.*, Jan. 1870.

23. A mass originally at rest is acted on by a force which in 1·368th of a sec. gives to it a velocity of $5\frac{1}{4}$ in. per sec.; show what proportion the force bears to the weight of the mass.—*Ib.*, Jan. 1871.

24 If a particle moves in consequence of the continued action upon it of a constant force, show what is the character of the resulting motion, and in what manner it depends on the magnitude of the force and the mass of the particle.—*Matriculation*, Jan. 1872.

25. As a special case show how the resulting motion would be changed if the mass of the particle were trebled, and the intensity of the force acting upon it were doubled. —*Ib.*, Jan. 1872.

26. The speed of a railway train increases uniformly for the first 3 mins. after starting, and during this time it travels 1 mile. What speed (in miles per hour) has it now gained, and what space did it describe in the first 2 mins.?—*Ib.*, June 1872.

27. In the last question, supposing the line level and disregarding friction and the resistance of the air, compare the force exerted by the engine with the weight of the train.—*Ib.*, June 1872.

28. Which could you throw further, a small ball of lead or a ball of cork of the same size? and why?—*Ib.*, Jan. 1873.

29. Find the tension on a rope which draws a carriage of 8 tons weight up a smooth incline of 1 in 5, and causes an increase of velocity of 3 ft. per sec.—*Ib.*, June 1873.

30. If, on the same incline, the rope breaks when the carriage has a velocity of 48·3 ft. per sec., how far will the carriage continue to move up the incline?—*Ib.*, June 1873.

31. When a body changes its rate of motion under the action of a constant force, show that the space described in any time is the same as the space described by a body moving uniformly with the mean velocity for the same time.—*Ib.*, Jan. 1874.

32. I suddenly jump off a platform with a 20 lb. weight in my hand. What will be the pressure of the weight

upon my arm while I am in the air? Give a reason for your reply.—*Matriculation*, Jan. 1874.

33. In Atwood's machine one of the boxes is heavier than the other by half an ounce. What must be the load of each in order that the overweighted box may fall through 1 foot during the first second?—*Ib.*, Jan. 1874.

34. The last carriage of a railway train gets loose whilst the train is running at the rate of 30 miles an hour up an incline of 1 in 150. Supposing the effect of friction upon the motion of the carriage to be equivalent to a uniformly retarding force equal to $\frac{1}{300}$ the weight of the carriage, find, first, the length of time during which the carriage will continue running up the incline, and secondly, the velocity with which it will be running down after the lapse of twice this interval from the instant of its getting loose.—*Preliminary Scient.* 1*st M. B.*, 1869.

35. A weight of 6 ozs. is drawn up along the lid, 4 ft. long, and rising 2 in 9, of a smooth desk by a weight of 5 ozs. which, attached to the other weight by a string, hangs over the top of the desk and descends vertically. Find the velocity acquired when the heavier weight reaches the top of the desk.—*Ib.*, 1871.

36. If the mass 1 lb. be the unit of mass, and 1 foot and 1 sec. be the units of space and time, how would you define the unit of force? and how many such units of force are there in the weight of $1\frac{1}{2}$ lbs.?—*Ib.*, 1871.

37. Two planes, inclined at one-third and two-thirds of a right angle to the horizon respectively, meet at the top and slope opposite ways. If bodies of equal weight fall down these planes, starting at the same instant, find their accelerations, and show that the motion of their centre of gravity is the same as if one of them were to remain at rest and the other to fall vertically. —*Ib.*, 1873.

38. The weights at the extremities of a string which passes over the pulley of an Atwood's machine are 500 and 502 grammes. The larger weight is allowed to descend; and 3 secs. after motion has begun, 3 grammes are removed from the descending weight. What time will elapse before the weights are again at rest?—*Prelim. Scient.* 1*st M. B.*, 1874.

CHAPTER IV.

NEWTON'S LAWS OF MOTION.

IX. *The First and Second Laws.*

§ 55. Three laws have been enunciated by Newton, as the principles in accordance with which motion takes place, and have generally been made the basis of Dynamical Science. In the preceding pages we have assumed some of the principles involved in these laws, but there are others of great importance which we shall now be able to consider. These laws are the following:—

§ 56. **Law I.**—*Every body continues in a state of rest, or of uniform motion in a straight line, unless it be compelled by impressed forces to change that state.*

This law, as founded on experience, is supported by negative evidence only. It consists of two distinct parts. The first part, which is sometimes called the Law of Inertia, asserts that a body has no power to put itself in motion, but if in a state of rest it will continue so, unless made to move by external force. Now, we have already seen that absolute rest nowhere exists, and that what we call rest is a complex state resulting from the combined action of several forces. Since there is no matter without force and no force without the tendency to motion, the state of rest cannot be regarded as the normal condition of bodies. What this law, therefore, asserts is, that when a body is maintained in a state of rest by the combined action of two or more forces, the energy of these forces (a term we shall afterwards more carefully consider) will be continuously preserved, and the body will consequently remain in a state of rest, unless some other force change the conditions.

The second part of the law asserts what we have already been compelled to assume, that all bodies tend [1] to move uniformly, and in a straight line. In other words, that a body once in motion tends to preserve that condition, and to maintain its original momentum. In support of this law, which observa-

[1] By introducing some word signifying 'tendency' in the propositions of dynamics, we save the necessity of providing for cases in which other forces change the motion of the body. Thus it is incorrect to say that all bodies fall to the earth, because a balloon contradicts this law; but it is perfectly correct to say that all bodies *tend* to fall to the earth, and to this law the motion of the balloon forms no exception.

tion abundantly verifies, an appeal is made to the experience acquired by gradually removing the several external causes, which practically, to a greater or less extent, always impede motion. Thus it is found that the smoother the road along which a stone is rolled the further the stone will roll; that the more the air is removed from a chamber in which a pendulum is suspended, the longer it will continue to oscillate;—that in all cases where a body once in motion is gradually brought to rest, some external force, such as friction, is shown to exist, and that as this force is diminished the duration of the motion is increased. This law shows not only that the velocity will be preserved, but likewise the direction of the motion; and that a body cannot move otherwise than in a straight line except by the continuous action of some external cause.

§ 57. **Law II.**—*Change of motion is proportional to the impressed force, and takes place in the direction of the straight line in which the force is impressed.*

This law asserts that whatever motion (and by motion is here understood quantity of motion or momentum) a certain force produces, double as much motion will be produced by twice the force, three times as much by a force three times as great, and so on. It also shows that if a certain force acting for one second can produce a certain amount of motion, the same force acting for two seconds will produce twice that quantity. This explains how it is that a constant force such as gravity produces an accelerative effect, adding on a fixed increment of momentum every second to the falling body. This law is really the fundamental principle of mechanical science: for since change of motion is proportional to the moving force, when several forces act together, the change of motion due to each is proportional to each, and the law of the composition of mechanical forces is thus established. This law may be enunciated thus:—

When several forces act simultaneously on a body, each produces the same effect as if it had acted separately.

The operations of this law have been already considered in the chapter on Kinematics; but quantity of motion was there understood to mean velocity only, since the mass of the body was not taken into account. This law includes, therefore, the law of the composition of velocities already referred to (§ 22). A consequence of this law is that forces produce the same effect when acting on a body at rest or in motion, since the state of rest is the result of a certain number of forces simultaneously acting. It also follows that the resultant of any number of forces may be found by the same process as the resultant of any number of velocities, and many of the propositions of Statics are immediately deducible therefrom. The law is illustrated by a variety of phenomena. A ball thrown vertically upwards from the hand of a person in motion will return to him just the same as if he had been stationary: a stone dropped from the top of the mast of a ship falls at the foot of the mast, whether the vessel be sailing or at rest: a body projected horizontally from the top of a cliff reaches the ground at the same point as if it had first moved horizontally with the velocity of projection, and then vertically downwards under the action of gravity for the whole time of flight. Other consequences of this law have been already considered in Chapter I.

X.—*The Third Law of Motion.*

§ 58. **Law III.**—*Action and Reaction are equal and opposite; that is, to every action there is a corresponding reaction equal in magnitude and opposite in direction.*

This law is a further development of the great physical principle involved in the two former laws—that Energy is as indestructible as Matter, and that when the action of a force seems to cease, we find in its place a new force which we call reaction. From this law many important principles, that properly belong to Statics, are at once deducible. It is illustrated by a variety of phenomena.

§ 59. **Statical Reaction.**—If one body presses another it is at the same time equally pressed by the other body, but in an opposite direction. If we press our hand on a table, the hand is equally pressed by the table, and if the table move under the pressure it is because the pressure of the hand against the table is just, and only just, greater than the pressure of the table against the hand. If a heavy body rest on a hard surface it presses against that surface, and is, at the same time, equally pressed by the surface in the opposite direction. This opposing pressure of the hard surface is called *Statical Reaction* and is always perpendicular to the surface, and exactly equal to the pressure which the body causes in that direction. This is true whether a body rests on a surface that is horizontal or inclined at an angle to the horizon. In the former case the reaction is equal to the whole weight of the body; but in all other cases it is equal to the pressure exerted, *i.e.* to the part of the force which acts in a direction perpendicular to the plane.

§ 60. **Tension.**—If a horse draw a tram-car by means of a rope, the horse is drawn in the opposite direction by a force equal to that which he exerts through the rope. The rope in fact is stretched by two equal forces in opposite directions. This effort, so to speak, of the rope in the direction of the horse and in the direction of the car is called *tension.*

Tension is a force that acts equally towards both ends of a stretched string and at any point in it. The force necessary to move a body when transmitted through a string must be just greater than the force of tension which acts equally towards the pulling body and the body pulled. That tension is a force which acts equally in both directions may be seen, if a string be attached to a fixed wall and stretched, and afterwards removed from the wall and stretched in the opposite direction by a force sufficient to keep the original force in equilibrium. These two forces will then be found to be equal, showing that the wall exerted a force equal to the stretching force. In speaking of the tension in strings, cords, &c. we shall assume that the string is perfectly flexible and inelastic; in which case the tension is the same throughout the whole length of the string. It acts at any part of it, towards both ends, and is equal to the stretching force.

§ 61. **Impact.**—If one body strike another body and change in any way the motion of the other body, its own motion will be changed in an equal quantity and in the opposite direction. Thus, if one body strike or impinge on another body the momentum which the one body loses is equal to that which the other gains. If the body A be moving in the direction XY, and the body B in the opposite direction, and if the momentum of A be equal to the momentum of B, and both bodies be perfectly inelastic, they will be brought to rest after they have impinged. But, if the momentum of A be greater than that of B, A will bring B to rest, and then both will move on with a diminished velocity. In these cases a certain

FIG. 18.

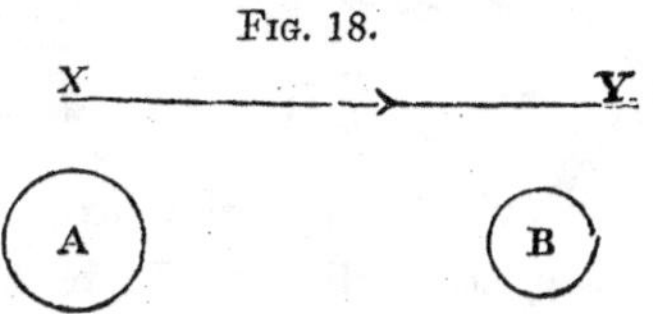

amount of momentum, and consequently of force, is apparently lost.

If the two bodies are moving in the same direction and A overtakes B, the two will move on with a new velocity which can be thus determined :—

Let the velocity of A be v and its mass M
and ,, ,, B ,, u ,, ,, N

Then the body A has Mv units of momentum
and ,, ,, B Nu ,, ,,

Let V be their common velocity after having impinged; then, since action and reaction are equal and opposite, the momentum lost by A equals the momentum gained by B, or

$$M(v - V) = N(V - u)$$

$$\therefore V = \frac{Mv + Nu}{M + N}.$$

If the bodies are moving in opposite directions

we can apply to their velocities the distinction of sign, which is common in the application of algebra to geometry, and call the velocity of one body positive and of the other negative. In this case

$$V = \frac{Mv - Nu}{M + N}$$

$$\therefore \text{ generally } V = \frac{Mv \pm Nu}{M + N}.$$

If the bodies be moving in opposite directions and $Mv = Nu$, $V = o$, *i.e.* the bodies will be brought to rest by their collision.

§ 62. **Examples:**

(1) Two bodies the masses of which are 5 lbs. and 9 lbs. are moving in the same direction with velocities of 10 ft. and 5 ft. respectively; what is their common velocity after impact?

Here $M = 5$, $N = 9$, $v = 10$, and $u = 5$

$$\therefore 5 \times 10 + 9 \times 5 = 14\,V \therefore V = \frac{95}{14} = 6\tfrac{11}{14}$$

ft. per sec.

(2) Two bodies whose masses are 10 lbs. and 8 lbs. are moving in opposite directions with velocities of 4 ft. and 6 ft. respectively; find the velocity and direction of the motion after impact.

Here $V = \dfrac{10 \times 4 - 8 \times 6}{18} = -\tfrac{4}{9}$ ft. per sec.

The motion is in direction of the lighter body.

§ 63. The third law of motion is further illustrated by the recoil of a gun, when a shot is projected from it. Without considering the exact action that takes place in consequence of the expansion of the gases which the ignited powder evolves, we may say that the momentum of the shot is found to equal the momentum of the gun.

Suppose a shot weighing 48 lbs. to start from a gun weighing 4 tons with a velocity of 200 ft. per second, the velocity of recoil can be easily determined For if v equal this velocity,

$$4 \times 20 \times 112 \times v = 48 \times 200,$$

and $v = 1\frac{1}{14}$ ft. per second.

EXAMINATION.

1. How does a body tend to move when once in motion? A stone is tied to a string and whirled round; if the stone free itself from the string in what direction will it move? Show how its motion illustrates Newton's first law.
2. Mention facts which serve to verify Newton's first law of motion.
3. A balloon is moving horizontally through the air at the rate of 30 miles an hour, and a stone dropped from it reaches the ground in 4 secs. What is the height of the balloon, and how far has it travelled during the passage of the ball? If the resistance of the air be considered, in what direction would the stone seem to fall?

4. What is meant by statical tension? How does it illustrate Newton's third law?
5. If two weights P and Q hang by a string over a smooth wheel, and P be greater than Q, what is the tension in the string?
6. Two bodies, perfectly inelastic, of different masses, are moving towards each other with velocities of 10 ft. per sec. and 12 ft. per sec. respectively, and continue to move after impact with a velocity of 1·2 ft. per sec. in the direction of the greater. Compare their masses.
7. What is the law of the composition of forces? How was it enunciated by Newton?
8. If forces of 30 lbs. and 40 lbs. act at a point at right angles to each other, what is the resultant force?
9. Two bodies of equal masses are moving in the same direction, and the velocity of the one is 10 ft. per sec., of the other 15 ft. per sec.; find their velocity after impact.
10. Two bodies whose masses are to one another as 3 : 2 are moving (1) in the same direction with velocities of 20 and 25 respectively, (2) in opposite directions; compare their velocities after impact.
11. A body the mass of which is 10 lbs. is projected along a smooth horizontal plane with a velocity of 20 ft., and strikes another body at rest the mass of which is 40 lbs.; find the velocity with which they move on together.
12. Three bodies of equal masses are placed at equal distances in a straight line on a smooth horizontal plane, a fourth body of equal mass is projected in the same line with a velocity of v ft. per sec.; find the velocities after successive impacts.
13. Show the application of Newton's third law, in hammering a nail into the wall, in the fall of a heavy body to the earth, in the driving of piles into the ground.
14. If a shot weighing 20 lbs. leave a gun weighing 3 tons

with a velocity of 1,200 ft. per sec., find the velocity of the gun's recoil.

15. If a shot weighing 32 lbs. be fired from a gun weighing 2 tons with a velocity of 1,120 ft. per sec., and if the friction between the gun and the ground be equal to a force of 1 ton, how far will the gun recoil?
16. Explain the terms 'force,' 'weight,' 'pressure,' and 'tension.'—*Matriculation*, Jan. 1869.
17. Distinguish between the statical and dynamical measures of force. How are they related to one another?—*Ib.*, June 1869.
18. Explain the third law of motion. Apply it to determine the velocity gained per second when weights of 6 ozs. and 4 ozs. are attached to the two ends of a string passing over the edge of a smooth table, the larger weight being drawn along the table by the smaller, which descends vertically.—*Ib.*, Jan. 1871.
19. State the three laws of motion, and give examples of each. A rifle is pointed horizontally with its barrel 5 ft. above a lake. When discharged the ball is found to strike the water 400 feet off. Find approximately the velocity of the ball.—*Ib.*, June 1873.
20. State the third law of motion.

    Show from your statement of it how to find the tension of the string and the acceleration when one ball is drawn up an inclined plane by another which hangs by a string passing over a fixed pulley at the top of the plane.—*Preliminary Scientific 1st M. B.*, 1870.
21. Equal spherical inelastic bodies are placed at short equal intervals in a smooth horizontal groove. The first is projected from an end along the groove with a velocity of 20 ft. per sec. Find the velocities after successive impacts.—*Ib.*, 1870.
22. Explain the meaning of the phrase 'Action and Reaction are equal and opposite.'—*Ib.*, 1872.
23. State the third law of motion, and show how it holds in the case of a stone which is in the act of falling towards the earth.—*Ib.*, 1873.

CHAPTER V.

ENERGY.

XI. *Work—Friction.*

§ 64. Whenever a body moves through any space in a direction opposite to that in which a force is acting on it *work* is said to be performed. If a horse draw a cart along a rough, level road, the horse does a certain amount of work, which depends on the resistance against which the horse is moving and on the space traversed. It is evident that the application of force is necessary to overcome resistance, and it is very often found convenient to measure the work done by the amount of force expended and the distance in the direction of the force, through which it has been employed. Thus, in the example already given, the force which the horse exerts, and the distance through which he moves, may be regarded as the two elements of the work done.

§ 65. **Unit of Work.**—The unit of work is that amount of work which is done when a body moves through one foot against a resistance of one pound. This unit of work is called the *foot-pound*. The unit of work is very commonly taken as the *kilogrammemetre*, and this unit is generally employed in comparing the work done by different physical agents. The most simple instance of the expenditure of a unit of work is when a pound-weight is raised vertically through one foot. In this case the measure of the force of gravitation is one pound, and the distance traversed is one foot, and we can fix our attention either on the force which is continuously exerted through one foot to raise the body, or on the resistance of one pound against which the body is moved through a foot. The result is the same in either case: a unit of work is performed.

§ 66. **Measure of work done.**—If a body move through s ft. against a resistance of R lbs., or be moved through s ft. in the straight line in which a force of R lbs. is acting on it, $R.s$ units of work will be performed. The work done, therefore, is measured by the product of the resistance or force, and the *effective distance* through which the body is moved. By effective distance is meant the distance measured in the same direction as that in which the force is acting. Thus, if a man pull a weight of 100 lbs. up a smooth incline 200 ft. in length and 80 ft. in height, the *effective distance* through which the weight will have been pulled is 80 ft., although the distance traversed is 200 ft., and the work done is 8,000 foot-pounds.

Friction.

§ 67. **Friction.**—The resistance which a force has to overcome in moving a body may be due to various causes, one of which, viz., gravitation, has already been considered. Another, quite as common, is friction, which is brought into play, whenever one rough body is moved upon another. Practically, all bodies are rough; for, although there exists hardly any limit to the degree of polish that can be imparted to certain substances, still no surface can ever be absolutely smooth.

§ 68. **Measure of Friction.**—It is evident that the resistance of friction acts in the opposite direction

Fig. 19.

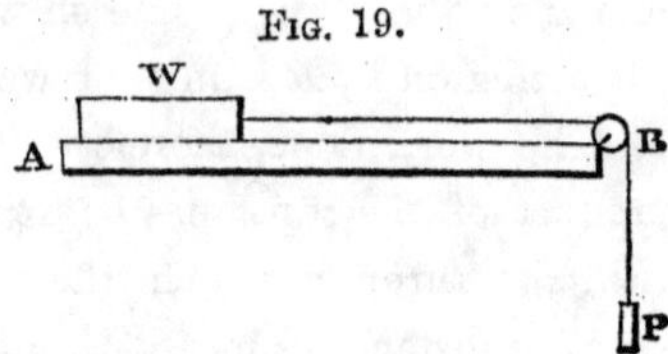

to that in which the body tends to move. It can be measured, therefore, by the force that is necessary to move the body on a horizontal plane. Suppose we require to know the resistance of friction that exists between the weight *W* and the surface *A B* on which it rests. Let a fine thread be attached to the weight and passed over a small wheel, as in fig. 19, and let the other end of the thread support a weight P. If P be the least weight which will cause the body to move along the plane P is the measure of the friction between the two surfaces.

§ 69. Another method of measuring the resistance due to friction is by elevating the plane $A\ B$

FIG. 20.

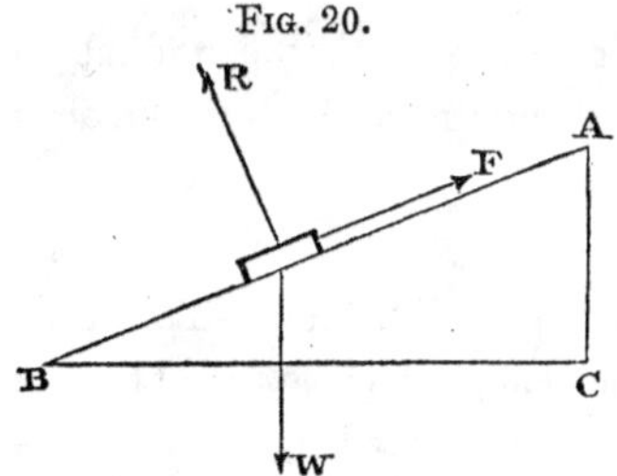

till the body begins to slide. If $A\ B\ C$ be the angle of elevation of the plane at which sliding commences, the resistance of friction is equal to that force which, acting parallel to the plane, would support the weight W, the plane being perfectly smooth. If F be this force $F : W :: h : l$ where h and l are the height and length of the plane (§ 52, Ex. 11) $\therefore F = \frac{h}{l} W$.

If R be the resistance of the surface it can be shown that $R : W :: b : l$, and hence

$$F = \frac{h}{l} \cdot \frac{l}{b} R = \frac{h}{b} R.$$

§ 70. **Laws of Friction.**—By these means the friction between different substances has been determined, and the following laws have been experimentally established:—

1. The friction is independent of the extent of surfaces in contact.

2. The friction is independent of the velocity when there is sliding motion.

3. The friction varies with the normal pressure, when the materials of the surfaces in contact remain the same.

§ 71. **Coefficient of Friction.**—The ratio of the friction to the normal pressure is called the *coefficient of friction*, and is represented by the Greek letter μ. Thus, if F be the force parallel to the surfaces in contact, which is just sufficient to produce sliding motion, and if R be the normal pressure $\frac{F}{R} = \mu$, or $F = \mu R$.

By normal pressure is understood the reaction of the surface, which we have seen (§ 59) is perpendicular to the surface pressed. Thus, if a weight rest on a horizontal surface the normal pressure equals the weight, or $F = \mu W$; but if it rest on an inclined plane, as in fig. 20, $F = \mu R = \mu \frac{b}{l} W$.

It follows from § 69 that the coefficient of friction

is the ratio between the height and the base of the plane on which sliding is about to commence.

§ 72. **Examples.**—(1) Find the work done in drawing a body of weight W up a rough inclined plane the height of which is h, and length l.

Let μ be the coefficient of friction. The work done against gravity is Wh, since the weight W is moved to a height h. The work done against friction is Fl, where F is the resistance due to friction. Now, we have seen that $F = \mu \frac{b}{l} W$ where b is the base of the plane. Therefore work done is $\mu b W$; and total work done equals $Wh + \mu b W$, or the same amount as would have been done if W had been moved along the base of the plane horizontally against friction, and then elevated against gravity through a height h.

(2) How much work is done when an engine weighing 10 tons moves half a mile on a horizontal road, if total resistance is 8 lbs. per ton?

The force against which motion takes place is 80 lbs., and the distance moved through is 2,640 ft.

$$\therefore \text{work done} = 211{,}200 \text{ foot-pounds.}$$

(3) If a weight of half a ton be lifted up by 20 men, 20 ft. high, twice in a minute, how much work does each man do per hour?

Here the weight lifted is 10 cwt. = 1,120 lbs., and the distance through which it is raised is 20 ft. ∴ the number of foot-pounds per minute is 44,800, and the number of foot-pounds per hour is 2,688,000, ∴ each man does 1,344,000 foot-pounds per hour.

(4) A heavy body falls down the whole length of a rough incline on which the coefficient of friction is 0·2. The height of the plane is 10 ft. and the base 30 ft. On reaching the bottom it rolls horizontally on a plane, having the same coefficient of friction. Find how far it will roll.

Now, in falling down the plane the body acquires, by the action of gravity, the power of doing a certain amount of work equivalent to what would be expended in raising the body to the top of the plane. Work acquired is $W h = 10\ W$ foot-pounds, where W is the weight of the body. Part of this number of units of work is lost by friction; and work expended in overcoming friction is $\mu\ b\ W = 0{\cdot}2 \times 30\ W = 6\ W$. Hence the body has acquired $4\ W$ units of work in falling. Now, if s be the distance through which the body can move along the horizontal plane against friction, the work expended will be $\mu\ W s$, $\therefore\ \mu\ W s = 4\ W$, or $0{\cdot}2 \times s = 4$ or $s = 20$ ft.

(5) A heavy body falls down a plane, as in the preceding Question, and with the energy acquired rolls up another plane in contact with it. Find the distance through which it ascends.

Let h, b, and μ be the height, base, and coefficient of friction on one plane. Let h', b', and μ' be the height, base, and coefficient of friction on the other. Then if W be the weight of the body, $W h - \mu\, W b$ is the number of units of work acquired in falling the whole length of the first plane. Let x be the distance measured horizontally, and y the distance measured vertically, from the foot of the plane, of the point which the body is enabled to reach on the incline.

Then, $W y + \mu' W x$ is the measure of the work expended in reaching this point, the positive sign

FIG. 21.

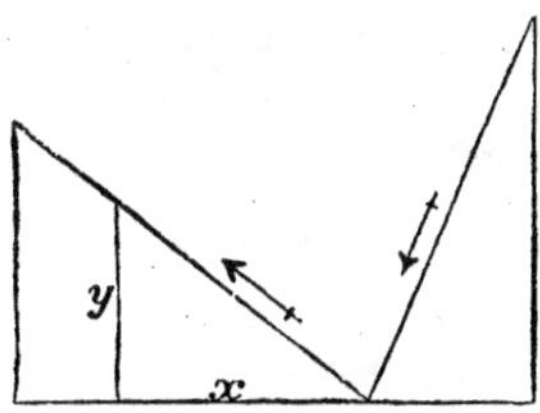

showing that gravitation and friction are opposed to the rising of the body.

$$\therefore W y + \mu' W x = W h - \mu\, W b,$$

or
$$y + \mu' x = h - \mu\, l,$$

also
$$\frac{y}{x} = \frac{h'}{b'} \therefore x\left(\frac{h'}{b'} + \mu'\right) = h - \mu\, b,$$

or
$$x = \frac{h - \mu\, b}{h' + \mu' b'} \cdot b'$$

and $$y = \frac{h - \mu b}{h' + \mu' b'} . h'$$

$$\therefore \text{Distance required} = \sqrt{(x^2 + y^2)} = \frac{h - \mu b}{h' + \mu' b'} \times l',$$

where l' is the length of the second plane.

XII. *Varieties of Energy—Conservation of Energy.*

§ 73. **Definition of Energy.**—The *energy* of a body is its power of doing work. Every moving body possesses energy, *i.e.* it is capable of overcoming resistance. Instances of bodies possessing energy are numberless. The flying bullet has the power of piercing a sheet of iron and of overcoming the cohesion between its particles; the running stream is able to turn the wheel of the water-mill, and the energy it possesses is utilised in grinding corn; the moving air drives the ship through the water and overcomes the resistance offered to its passage. Wherever we find matter in motion, be it solid, liquid, or gaseous, we have a certain amount of energy which can be utilised in endless ways.

§ 74. **Kinetic and Potential Energy.**—The energy of moving matter is called *Kinetic Energy*: it is energy ready for use,—energy that is constantly being spent, though it may not be economically employed. The rivulet will flow from the highlands to the lowlands, whether we give it work to do or not.

Now, the energy of a moving body can be conveniently expressed in terms of its rate of motion. Suppose v to be the velocity of the moving mass and W its weight, and suppose s to be the vertical height to which W would rise if projected upwards with the velocity v, then $W s$ is the measure of the work which W is capable of doing; and since $v^2 = 2\,g\,s$,

$$W s = \frac{W\,v^2}{2g} = \frac{M\,v^2}{2}.$$

Hence, the kinetic energy of a mass M moving with a velocity v is equal to $\frac{M\,v^2}{2}$, and this expression represents the number of units of work stored up in the moving body and available for any purpose to which the hand of man can adapt it.

But matter is not deprived of energy even when at rest. We have seen that bodies at rest have a tendency to motion, which the removal of a force, or the application of a force, can always render actual. In the same way there exists in so-called inert matter, —in matter that is apparently at rest, a tendency to put forth energy, and this energy is called *Potential Energy*. Suppose a weight of 1 lb. be projected vertically upwards with a velocity of 32·2 ft. per second. The energy imparted to the body will have carried it to a height of 16·1 ft., and the body will then cease to have any velocity. The whole of its kinetic energy will have been expended; but the body will have acquired, instead, a new position, a vantage-ground; and if free to fall from this position it will obtain a velocity of 32·2 ft. per second, and thus re-acquire the energy which it originally received. Now, we may suppose the weight to be lodged for any length of time at an elevation of 16·1 ft. from its point of projection, and during this time its energy will be *latent* or *potential*—stored up and ready to be freed, whenever it shall be permitted to fall. It is evident that the matter of which this earth is composed is always endeavouring to find a lower level than it possesses; and this tendency is the source of its potential energy. The energy, which in bygone years has been expended in raising walls and towers on the tops of hills and elsewhere, still survives, and when the stones of which they are composed shall fall from their places, there will be expended in the fall the same amount of energy as was employed in raising them. In the boulder embedded in the sea-shore we have evidence of kinetic energy that has been spent; in the overhanging crag we see a mass endowed with the energy of position, which at any moment may be changed into an active and destructive force. Nature, by her own processes, is continually modifying the relative position of the matter of which this earth is formed, and in every change there is a re-adjustment of energy, but neither loss nor gain.

§ 75. **Conversion of Heat into Mechanical Energy.**—Our chief sources of heat are the sun and the combustion of fuel. The sun's heat is being constantly changed into mechanical energy. The wind that turns the sails of the mill and drives the ship through the sea receives its motive power from the sun's heat. The mechanical energy of a running stream has been derived from the same source. The clouds that settle on the ridges of the hills and on the mountain-peaks have been raised by the force of the sun's heat from the waters of the sea; and the falling rain feeds the streams and rivulets which, after having done some useful work, find their way back to the sea. The havoc which so often follows the glacier in its fall is caused by the release of the sun's energy stored up in the ice.

When wood or coal is burned the heat evolved can be made to perform work. The steam-engine is the most important instrument for converting heat into mechanical energy. A portion of the energy thus formed is employed in overcoming certain resistances incidental to the working of the machine; the remainder may be spent directly in work, as in raising heavy loads by the steam-crane, or in giving motion to other bodies, as in the ordinary locomotive. In either case the heat is ultimately transmuted into work.

§ 76. **Conversion of Mechanical Energy into Heat.**—In discussing Newton's third law of motion we showed that when one inelastic body impinges on another, the momentum lost by one body is gained by the other, and that if they are moving in the same direction, the momentum of both, after impact, is the same as that of the two separately before impact. Moreover, when two inelastic bodies, each having the same momentum and moving towards each other, impinge, the momentum of the one destroys that of the other, and they come to rest. In both these cases, by employing the algebraical conventions of *sign*, to distinguish velocity in one direction from velocity in the opposite direction, we may say that the momentum is the same, before and after impact. But not so the kinetic energy. If we suppose two bodies the masses of which are 10 lbs. and 4 lbs. to be moving in the same direction, with velocities of 8 ft. and 15 ft. respectively, their kinetic energy before impact is $\frac{10 \times 8^2 + 4 \times 15^2}{2 \times 32} = 24\frac{1}{16}$; while that after impact is $\frac{10+4}{64} \times 10^2 = 21\frac{7}{8}$ (where 10 is their common velocity after impact, § 61). Thus there is an apparent loss of kinetic energy. If the two bodies are moving in opposite directions, or if an inelastic body in motion strike a similar body at rest, there will be a still greater loss of kinetic energy. In all these cases, however, the energy that is lost by impact reappears in the heat generated by the blow; and hence, by widening the signification of Newton's law, so as to embrace any form of energy, the truth of the law is maintained.

We have seen, that when one body moves on the surface of another body a resistance is offered to the motion, which is due to friction. If a body be projected along a rough road with a certain velocity, it will after a time come to rest, and the kinetic energy of the body will have been destroyed by friction. Now, in this case, a corresponding amount of heat will have been produced, exactly equivalent to the energy destroyed. One of the earliest ways of obtaining heat was by rubbing sticks together, in other words by evolving it from mechanical energy. We have seen that in a locomotive-engine heat is changed into kinetic energy; but the heat employed must be in excess of the kinetic energy required, as a large amount of friction has to be overcome, and this friction reproduces heat. Hence, part of the heat-energy, evolved from the coals, after having been transformed into mechanical energy, reappears in the increased temperature of the rails, axles, and other parts of the machinery. If the speed of the engine remain uniform, the heat of the furnace is wholly employed in overcoming these resistances, and is spread over the rails and machinery; and if the steam be turned off and brakes applied, the kinetic energy of the engine is gradually converted into heat, and the locomotive stops. Whenever a body is brought to rest by moving over a rough surface, by passing through water or air, or by striking against another body, the energy of the moving body is not destroyed, but is converted into a new form of energy, equal in amount to that which is apparently lost.

§ 76 (*a*). **Dissipation of Energy.**—In order that heat may be converted into mechanical energy, it is necessary that it should pass from a body of high temperature to one of low. This occurs in every heat-engine, which consists essentially of two chambers, 'one of high, and the other of low temperature,' and which performs work 'in the process of carrying heat from the chamber of high to that of low temperature.'[1] In the fall of temperature, as in the fall of water from a higher to a lower level, work is done, and a certain amount of available energy is lost. On the other hand, we have seen that whenever a body moves against friction a certain amount of heat is generated, as when a brass button is rubbed on a piece of dry wood; but the heat thus generated possesses so low a temperature as not to be available for the purposes of work. By no process, yet known, can this diffused heat be reconverted into mechanical energy. It thus appears that a large amount of

[1] B. Stewart's *Conservation of Energy*.

energy, though not absolutely lost, is always becoming practically useless, through the conversion of heat at a high, into heat at a low temperature. This loss of available energy takes place whenever work is evolved from fuel and other sources of heat at a high temperature; and, in the absence of all knowledge with respect to the means by which the sun's heat may be renewed, a theory of the Dissipation of Energy has been arrived at, which asserts that 'the mechanical energy of the universe will be more and more transformed into universally diffused heat, until the universe will be no longer a fit abode for living beings.'[1]

§ 77. **Relation between Work and Heat.**—A number of experiments have been made to show what amount of heat is exactly equivalent to a unit of work. The details of these experiments are given in treatises on Heat. They serve to prove that the quantity of heat necessary to raise the temperature of a pound of water through 1° F. possesses the same amount of energy as is required to lift 772 lbs. through one foot. In other words, the energy of a unit of heat is equivalent to 772 units of work. In comparing work with heat, the kilogramme-metre and degree Centigrade are generally taken as standards. In this case the unit of heat is the heat re-

[1] B. Stewart's *Conservation of Energy.*

quired to raise a kilogramme of water through 1° C., and the unit of work is the energy necessary to lift a kilogramme through a metre of height. The relation that subsists between these quantities is given in the statement: That a unit of heat can generate 424 units of work, and in the destruction of 424 units of work one unit of heat is produced.

§ 78. **Conservation of Energy.**—The foregoing are a few instances of a vast number of facts, which have been generalised of late years into a fundamental law of physical science. This law, known as that of the Conservation of Energy, embraces the following propositions:—

(1) The sum-total of energy in the universe remains the same.

(2) The various forms of energy may be converted the one into the other.

(3) No energy is ever lost.

§ 79. (1) These three propositions are intimately connected with one another. To understand the first of them, we must suppose the universe to have been originally endowed with certain energies, the sum-total of which is always, in quantity, the same. We have seen that various kinds of forces exist. The most important of these is Gravitation, with which matter in all its forms is universally endowed. It acts between bodies in mass and between the molecules of bodies. Other mechanical forces are Cohesion, which holds the particles of a body together, and Elasticity, which is exhibited in a watch-spring, and which causes the particles of a body to resume their original position after having undergone displacement. We have spoken of Heat as a force, and have shown the relation between mechanical and thermal energy. The remaining forces of Nature are less capable of being brought within the range of mechanical principles. These comprise Chemical-affinity, which unites the elements that constitute a molecule or particle, Light, Electricity, and Magnetism. With respect to these forces, we can do no more than indicate the names by which they are known. What the law of conservation asserts is, that the sum-total of energy due to all these forces always has remained and always will remain the same. If the energy due to any one force be diminished in quantity there must be a corresponding increase in some other variety of energy. It will be seen that this law of the totality of physical energy necessarily involves the second of the three propositions above stated.

§ 80. (2) **Transmutation of Energy.**—Facts have already been mentioned that illustrate this law. The most general form of energy is what has been called 'visible energy,' *i.e.* matter in motion. This can easily be converted into latent energy or energy of position. These two states are exemplified in the oscillation of a pendulum. When the pendulum reaches its lowest point it possesses a certain amount of kinetic energy, which is sufficient to raise it through a certain arc on the other side, and when it reaches its highest point it acquires an equivalent amount of latent energy, which being set free, is reconverted into energy of motion. This transmutation of kinetic into latent energy, and of latent into kinetic energy, might go on for ever, if the friction of the pivot and the resistance of the air did not gradually retard the motion of the pendulum. A certain amount of energy thus disappears, and we have found that the energy of motion that is lost is really converted into heat. The pivot and the air become heated by the motion. Kinetic energy, though it may endure for any limited period of time, must ultimately waste away into heat; and all moving bodies, in so far as they move against friction or through a resisting medium, will eventually come to rest. Perpetual motion, in the sense in which it is generally understood, is thus demonstrated to be impossible. Since, however, heat is shown to be a form of motion, it would appear that the motion of masses is perpetually changing into the motion of molecules.

The connection between heat and motion is better understood than the relation between other forms of energy; but the researches of modern science are continually showing us how other varieties of energy are capable of being transmuted and reproduced under new forms. 'The exact nature of the various kinds of molecular energy, such as heat, light, electricity, magnetism, and chemical affinity, is not at present known, but we run little risk of error in affirming that they all consist either of peculiar kinds of molecular motions, or of peculiar arrangements of molecules as regards relative position. They must, therefore, fall under one or other of the two heads, "Energy of Position" and "Energy of Motion."'[1]

§ 81. (3) **The Indestructibility of Energy.**—Our law asserts that energy is never lost. It may become latent and remain so for centuries; but it is permanent. The great source of energy in this universe is the sun. Streams of energy are continually flowing from the sun to the earth, and this energy is made to do all kinds of useful work. It gives to the waters of our seas an energy of position, by converting them into vapour and raising them to the hill-tops. It supplies the vegetable world with the power of absorbing carbon from the air, and the carbon thus fixed in the plant is gradually converted into wood, which may be immediately employed as fuel, or may reappear as coal after having been

[1] Deschanel's *Natural Philosophy*, translated by Everett.

buried in the earth for thousands of years. In either case the sun's energy, stored up in the fuel, is eventually set free to be converted into work. The vegetable products of the earth supply food to animals, and these again supply food to man; and thus the energy of animal agents is ultimately derivable from the sun. Our scientific knowledge may be too limited to enable us, in all cases, to trace the course of energy through its various transmutations; but every new observation that is made brings with it additional evidence tending to verify the law that energy is never lost.

§ 81 (*a*).—**Relation between Force, Momentum, and Energy.** From what has gone before, it appears that a body can only be brought to rest by the action of a force or resistance in a direction opposite to its motion. If we wish to estimate the amount of force that must be expended in order to reduce the body to rest, we may proceed in two ways, according as we wish to consider the *time* of motion, or the *space traversed* against the resistance. In the former case the force may be measured by the momentum of the body, in the latter by the energy. We have seen that where P is a force, which acting on M units of mass for one second gives it a velocity f, $P = M f$; and, if the moving body has a velocity v, which will cause it to move for t seconds against P, then $P = \frac{M v}{t}$, or $P t = M v$. Similarly, if we wish to consider the space s, through which the body will move in virtue of its velocity v, v and s are related by the equation $v^2 = 2fs$, and $\therefore P = \frac{Mv^2}{2s}$, or $Ps = \frac{Mv^2}{2}$. Hence, the product of the force into the *time* of motion represents the momentum, and the product of the force into the *space traversed* the energy of the moving body. By letting P, t, and s each equal unity in the two equations $Pt = Mv$ and $Ps = \frac{Mv^2}{2}$, we see that the unit of *momentum* is that amount of momentum which is produced by a unit of force, acting for a unit of *time*, and that the unit of *energy* is produced by a unit of force acting through a unit of *space*.

Thus, suppose, as in problem (13) § 52, it is required to find the *distance* up a plane which a body will ascend, when the velocity of projection is 40 ft. per sec., and the resistance due to friction $\frac{1}{100}$ the weight, the plane rising 1 in 10.

Here, the energy of the projecting force is

$$\frac{Mv^2}{2} = \frac{Wv^2}{2g} = \frac{W \times 1600}{2g};$$

and this energy is destroyed by the work done against friction and gravity. Hence—

$$\frac{W \times 1600}{2g} = \frac{W}{10}.s + \frac{W}{100}.s$$

$$\therefore s = \frac{1600}{2g} \times \frac{100}{11} = 227\tfrac{3}{11} \text{ ft.}$$

EXERCISES.

1. A body is just on the point of sliding on a rough plane that rises 3 in a length of 5; find the coefficient of friction.
2. What is the tension in a string which, being stretched, is just able to move a body weighing 10 ozs. over a rough horizontal plane, coefficient of friction being $\frac{2}{5}$?
3. A rough plane rises 3 in 10, and a body weighing 12 ozs. is just supported by friction; find the force of friction.
4. Find the work expended in pulling a body weighing 3 cwts. 100 yards up an incline that rises 1 in 10, if the force of friction be 10 lbs. per cwt.
5. A body weighing 12 ozs. is partly supported on a rough inclined plane that rises 1 in 2 by the force of friction and partly by a force parallel to the plane; if the plane be lowered so as to rise 2 in 5, the force of friction alone will support it; find the additional force in the former case.
6. A train weighing 10 tons moves for 20 minutes at a uniform rate of 30 miles an hour; if the friction, &c. be 10 lbs. per ton, how much work is done during the time
7. How much work is done per hour if 100 lbs. be raised 3 ft. in 1 minute?
8. If a pit 10 ft. deep and with an area of 4 square ft. be excavated, and the earth thrown up, how much work will have been done, supposing a cubic foot of earth to weigh 90 lbs.?
9. A body weighing 40 lbs. is projected along a rough horizontal plane with a velocity of 150 ft. per sec.; the coefficient of friction is $\frac{1}{8}$; find the work done against friction in five seconds.
10. What amount of energy is acquired by a body weighing 30 lbs. that falls through the whole length of a rough inclined plane, the height of which is 30 ft., and the base 100 ft., the coefficient of friction being $\frac{1}{5}$?

EXAMINATION.

1. Define a unit of work.
2. Given the weight and velocity of a moving body, find an expression for its kinetic energy.
3. Distinguish between energy of motion and energy of position.
4. What are the laws of friction? How may the coefficient of friction for different substances be determined?
5. A body is projected up a rough inclined plane that rises 7 to a base of 10 with a velocity of 200 ft. per sec.; find its velocity when it returns to the point whence it was projected, the coefficient of friction being 0·3
6. A body weighing 50 lbs. is projected along a rough horizontal plane with a velocity of 40 yards per sec.; what amount of work is expended when the body comes to rest, and what is the equivalent of heat generated?
7. A body whose weight is W is moving with a velocity V, and afterwards is found to move with a velocity u; give an expression for the energy lost.
8. Illustrate by the impact of bodies the law of the Conservation of Energy.
9. What is the origin of the energy possessed by coal? When coal is consumed by a locomotive-engine, what becomes of its energy?
10. If $g = 9{\cdot}8$ metres, find the equivalent of heat necessary to project 50 kilogrammes with a velocity of 490 metres per second.
11. Force may be defined as 'any cause which tends to change a body's state of rest or of uniform motion in a straight line.' Mention any forces you know of, and show how this definition applies to them. What is inertia? Is it a force?—*Matriculation*, 1872.
12. When a particle which is acted upon by any number of forces moves during their action with uniform velocity

in a straight line, show what condition the forces must fulfil.—*Matriculation*, 1872.

13. How is the energy of a moving body estimated? Through what distance must a force equal to the weight of $\frac{1}{2}$ lb. act upon a mass of 48·3 lbs. in order to increase the velocity from 24 ft. to 36 ft. per second?—*Prelim. Scient. Exam.*, 1869.

CHAPTER VI.

MACHINES.

XIII. *General Remarks—Application of the Law of Energy.*

§ 82. A machine is an instrument by means of which a force applied at one point is able to exert, at some other point, a force differing in direction and intensity. The force applied is called *the power*, the force exerted, or effectual resistance overcome, is called *the weight.* The resistance to be overcome may be the earth's attraction, as in raising a weight; the molecular attractions between the particles of a body, as in stamping metal, or dividing wood; or friction, as in drawing a heavy body along a rough road.

§ 83. Besides the *weight*, or effectual resistance, the power is employed in overcoming the internal resistances, chiefly due to friction, which always exist between the different parts of a machine. The power may be just sufficient to overcome these two kinds of resistance; it may be in excess of what is necessary, or it may be too small. If just sufficient, the machine once in motion will remain uniformly so, or if it be at rest it will be on the point of moving, and the power, weight, and friction will be in equilibrium. If the power be in excess, the machine will be set in motion and will continue to move with an accelerated motion; if the power be too small, it will not be able to move the machine; and if the machine be already in motion, it will gradually come to rest. When the power is just sufficient to overcome the weight, the ratio of the weight to the power is called the *modulus* of the machine. In this case it is evident, from the law of Conservation of Energy, that the work done by the power has its equivalent in the work done against the resistance to be overcome and against the friction between the parts. This is the general law of machines in equilibrium or in uniform motion, and may be formulated thus:—Let P be the power employed, p the distance through which it moves in a given time in its own direction, W the external resistance overcome, and w the distance through which it is moved in the same time; let F be the force of friction and other internal resistances, f the space through which these act; then work done by P equals work done against W and F,

or $$Pp = Ww + Ff.$$

This law shows us that whilst P can be made as

small as we please by taking p great enough, the mechanical advantage of diminishing P is restricted by the fact that f increases with p; and consequently with the decrease of P there is a corresponding increase of the work to be done against friction. Hence, if friction be neglected, there is no practical limit to the ratio of P to W; but if the friction between the parts be considered, the advantage of decreasing P has a limit, since if Pp remains the same, Ww must decrease as Ff increases; in other words, the work done against friction increases with the complexity of the machinery.

§ 84. In the application of this principle to those simple machines generally known as the *Mechanical Powers*, we shall not take into consideration the resistance due to friction, but shall suppose $F = O$, in which case the law of the machine becomes $Pp = Ww$, or $\frac{W}{P} = \frac{p}{w}$. This equation shows that the ratio of the *weight* to the *power* is equal to the ratio between the distance through which the power moves, and the distance through which the weight is moved, in the same time. Or, if we take equal distances and uniform motion, we see that *what is gained in power is lost in time.* When W is greater than P the machine is said to work at a mechanical advantage, when less at a disadvantage; and the ratio $\frac{W}{P}$ when greater than unity, is called the *me-*

chanical advantage of the machine. Machines are frequently employed for enabling a large power to move a smaller weight through a greater space than the power itself moves. With respect to the force applied, such machines work at a mechanical disadvantage; but, with respect to the space traversed, they work at an advantage. The advantage or disadvantage of a machine depends, therefore, on the object to be gained. The steam-engine and the watch are both machines in which the power expended exceeds the resistance to be overcome, but the equation of work holds good.

When a machine is in uniform motion $Pp = Ww$, or $Pp - Ww = 0$; i.e. the algebraical sum of the works done by all the forces is zero. If the machine be in equilibrium the principle of work equally holds good. For, if we suppose the machine to undergo a small displacement, consistent with the connection of its several parts, then, since the forces balance one another, the algebraical sum of the works of the forces must be zero. In this case, the velocities, being imaginary, are called *virtual*; and the principle is known as the principle of *virtual velocities*, and may be enunciated thus: If any machine is in equilibrium under the action of forces, and we suppose it subjected to any small displacement, consistent with the connection of its parts, the algebraical sum of the works done by all the forces is zero; and, conversely, if the sum be zero, the forces are in equilibrium.

§ 85. **Simple Machines.**—The simple machines, sometimes called the *Mechanical Powers*, may be considered under three heads, according as they involve:

I. A solid body moveable about a fixed point.
II. A flexible string.
III. A hard inclined surface.

Under the first head are comprised (1) the *lever* and (2) the *wheel and axle*; under the seeond head (3), the various kinds of *pulleys*; under the third head (4), *the inclined plane* (5), the *wedge*, and (6) the *screw*.

XIV. *Lever.—Wheel-and-Axle.*

§ 86. **The Lever.**—A rigid rod turning on a fixed point is called a lever. The fixed point is called the *fulcrum*, and those parts of the rod on either side of the fulcrum are called the *arms*.

Let $A\ B$ be a lever turning freely about C, the fulcrum, and let P be the force applied at A, and W the force exerted, or resistance overcome, or weight raised at B. Suppose the lever turned through the angle $A\ C\ A'$, then the work done by P equals $P \times$ arc $A\ A'$, and work done by W equals

FIG. 22.

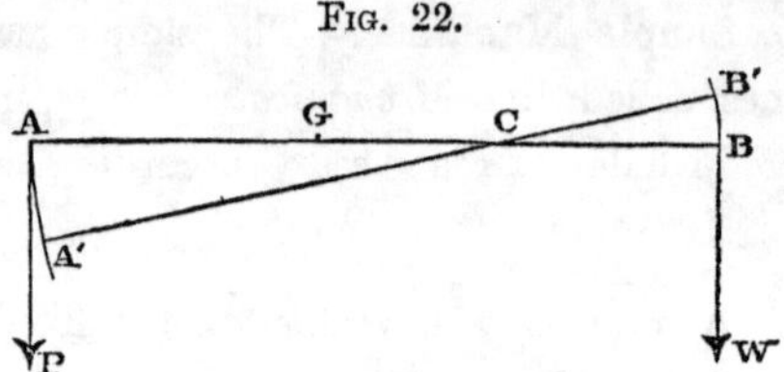

$W \times$ arc $B\ B'$, if P and W act perpendicularly to the arm. Therefore, by the law of energy, $P \times A\,A' = W \times B\,B'$

and since $\frac{A\ A'}{B\ B'} = \frac{A\ C}{B\ C}$ we have $P \times A\ C = W \times B\ C$,

or $P \times$ its arm $= W \times$ its arm.

This is the general principle of the lever, when the bar is a straight rod and the forces are perpendicular to it.

§ 87. **Three kinds of Levers.**—Levers are of three kinds, according to the position of the fulcrum with respect to the power and the weight.

Where the fulcrum is in the middle the lever is of the first kind; where the weight is in the middle, the lever is of the second kind; where the power is in the middle, the lever is of the third kind.

In each of these three kinds

$$\frac{W}{P}=\frac{A\,C}{B\,C}.$$

In I. there is a mechanical advantage or disadvantage, as $A\ C$ is greater or less than $B\ C$.

Fig. 23.

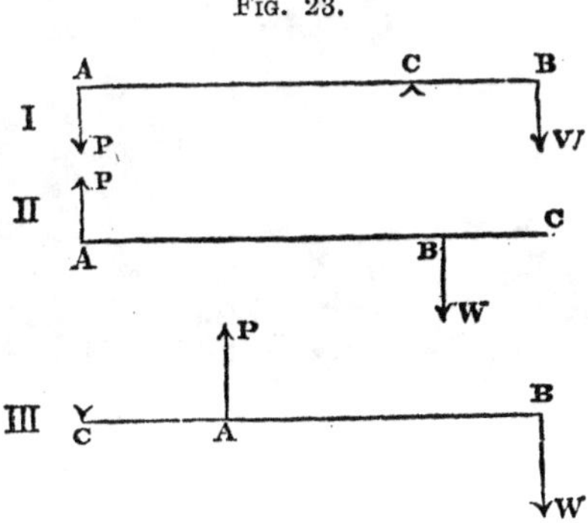

In II. $A\ C$ is necessarily greater than $B\ C$, and there is always a mechanical advantage.

In III. $A\ C$ is necessarily less than $B\ C$, and there is always a mechanical disadvantage.

Besides the forces P and W, there is the reaction of the fixed point or fulcrum to be considered, which is equal to the pressure of the fulcrum on the bar.

In I. the reaction or pressure is equal to $P + W$.
In II. " " " $W - P$.
In III. " " " $P - W$.

By means of a lever two forces, however unequal, may be made to balance each other, by so adjusting the position of the fulcrum that its perpendicular distances from the directions of the forces shall have the inverse ratio of the forces.

§ 88. **Examples of Levers.**—The crowbar, fig. 24, is a good example of a lever of the first kind.

Fig. 24.

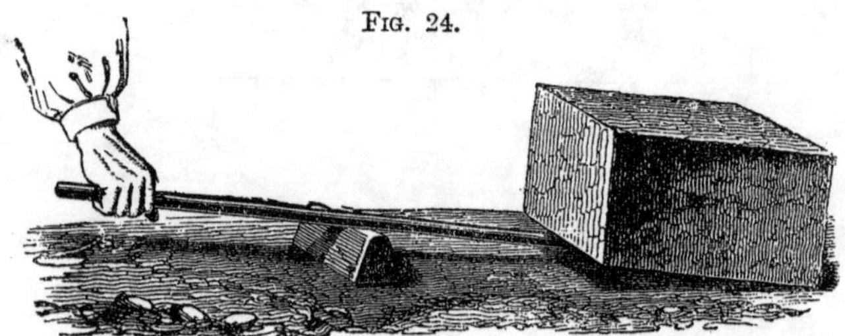

Fig. 25.

In figs. 25 and 26 we have examples of the second and third orders of lever. The wheelbarrow is a lever of the second kind, in which the fulcrum acts

Fig. 26.

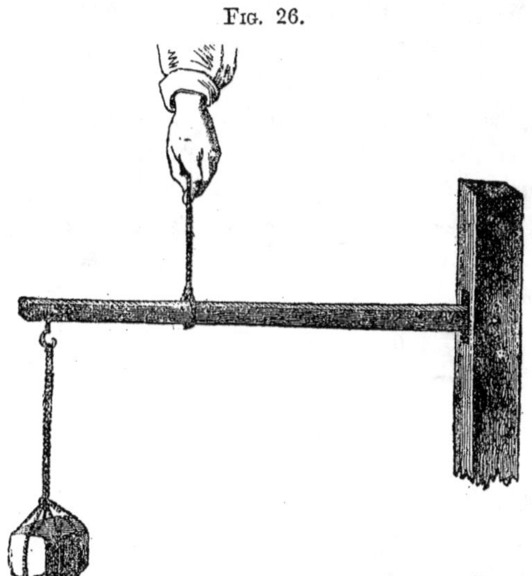

at the axle of the wheel and the power is applied at the handle. A rowing-boat is propelled by a lever of the second kind, in which the fulcrum is at the point where the oar enters the water, whilst the weight is the resistance of the boat fixed to the oar at the row-locks. The treadle of a turning-lathe is an example of a lever of the third kind. The best examples of this order of levers are seen in the animal frame, in which readiness of action is obtained at a loss of power.

§ 89. **Weight of the Lever.**—If the weight of the lever be taken into account, we have an additional force to consider, which may assist the action of the power or give the power more work to do, according as it acts on the same side of the fulcrum as the power or not. If G be the position of the centre of gravity of the lever (fig. 22) and w the weight of the lever acting at this point, then the equation of work becomes

$$P \times AC + w \times GC = W \times BC$$

or $P \times AC = W \times BC + w \times GC$, according to the position of G with respect to C.

§ 90. **Examples.**—(1) A lever is 3 ft. long, and its weight is 1 lb., and acts at the middle point. The resistance to be overcome at one end is 20 lbs., and the force applied at the other end is 3 lbs.; find the position of the fulcrum.

Let x be the distance of the fulcrum from A where P acts (fig. 22); then $3-x$ is the distance of the fulcrum from B where W acts, and $x-\frac{3}{2}$ is the distance of the fulcrum from G where w acts.

$$\therefore P \times x + w \times \left(x - \frac{3}{2}\right) = W \times (3 - x)$$

$$\therefore 3x + x - \frac{3}{2} = 20(3 - x) = 60 - 20x$$

$\therefore x = 2\frac{9}{16}$ ft. = the distance of fulcrum from A.

(2) What force must be applied at the end of a lever 20 inches long to raise a weight of 30 lbs. slung at a point 5 inches from the other end, if the weight of the lever is 4 lbs. and acts at its middle point?

Here $\quad P \times AC = 4 \times CG + 30 \times CB$

or $\quad 20P = 4 \times 10 + 30 \times 5$

$$\therefore P = 9\tfrac{1}{2} \text{ lbs.}$$

The various kinds of balances are examples of levers; but these will be treated later on, under the head of *Moments*, when the principle of levers will be further considered.

EXERCISES.

1. A lever is 18 ins. long; where must the fulcrum be placed in order that a weight of 6 lbs. at one end may balance double its weight at the ether end?
2. What force must be applied at one end of a lever 12 ins. long to raise a weight of 30 lbs. hanging 4 ins. from the fulcrum which is at the other end, and what is the pressure on the fulcrum?
3. Two weights of 6 lbs. and 8 lbs. are hung from the ends of a lever 7 ft. long; where must the fulcrum be placed so that they may balance?
4. A weight of 10 ozs. at the end of a lever is raised by a force which is just greater than 36 ozs., and which acts 6 ins. from the fulcrum which is at the other end;

what is the length of the lever and the pressure on the fulcrum?

5. A lever weighs 3 ozs., and its weight acts at its middle point; the ratio of its arms is 1 : 3. If a weight of 48 ozs. be hung from the end of the shorter arm, what weight must be suspended from the other end to prevent motion?

6. A lever 10 ins. long, the weight of which is 4 ozs., and acts at its middle point, balances about a certain point when a weight of 6 ozs. is hung from one end; find the point.

7. A heavy lever weighing 8 ozs. balances at a point 3 ins. from one end and 9 ins. from the other. Will it continue to balance about that point if equal weights be suspended from the extremities?

8. A beam the length of which is 12 ft. balances at a point 2 ft. from one end; but if a weight of 100 lbs. be hung from the other end, it balances at a point 2 ft. from that end; find the weight of the beam.

9. A beam the length of which is 8 ft. balances at a point 2 ft. from one end. If to this end a weight of 40 lbs. be hung, find the least force applied at the other end that will support this weight.

10. A heavy beam 16 ft. long, and weighing 4 lbs., balances by itself about a point 4 ft. from one end. If a weight of 10 lbs. be hung 2 ft. from this end, find the weight that must be hung from the other extremity that the beam may balance about its middle point.

---

§ 91. **Wheel-and-Axle.**—The wheel-and-axle is a modification of the lever. It consists of two cylinders having a common axis, the larger of which is called the wheel and the smaller the axle. The axis common to both is horizontal. A weight *W* hangs at the end of a string fastened to the axle and is coiled round it by the revolution of the wheel. The

Fig. 27.

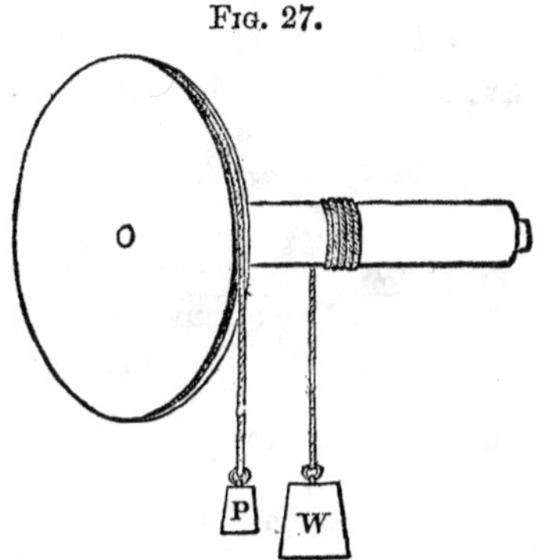

wheel may be moved by the hand or by a string with a weight attached to it.

Viewed in section we have the power *P* acting

Fig. 28.

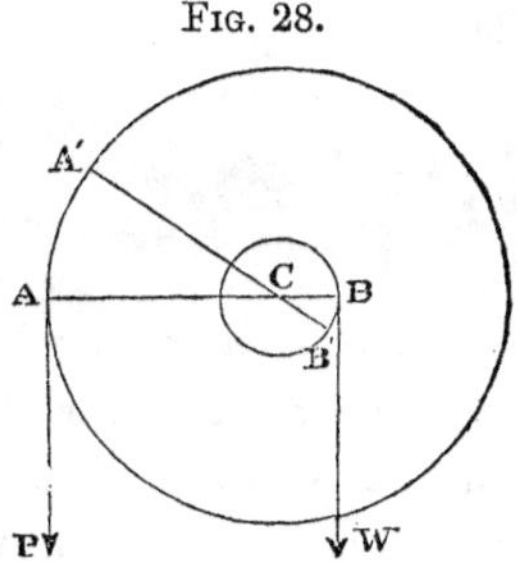

at the end of the radius of the large wheel at *A*, and the weight *W* at the end of the radius of the smaller wheel or axle. As P descends W ascends; and if the wheel move through an arc $A\,A'$, so that A' comes to A, B' will in the same time reach B. Then the work done by P is $P \times A\,A'$, and the work done by W is $W \times B\,B'$ $\therefore$ $P \times A\,A' = W \times B\,B'$, where P is just sufficient to overcome W,

or $$\frac{W}{P} = \frac{A\,A'}{B\,B'} = \frac{A\,C}{B\,C}$$

i.e. $$\frac{W}{P} = \frac{\text{radius of wheel}}{\text{radius of axle}} = \frac{\text{circumference of wheel}}{\text{circumference of axle}}.$$

§ 92. Examples of the wheel and axle are very numerous. One of the commonest is the windlass,

Fig. 29.

in which power is applied at the end of a handle. In the capstan (fig. 30.) the axle is vertical, and the mechanical advantage is increased by the length of the pole inserted into the circumference. The various kinds of mills are examples of this machine, the weight appearing in the form of a resistance to be overcome.

Fig. 30.

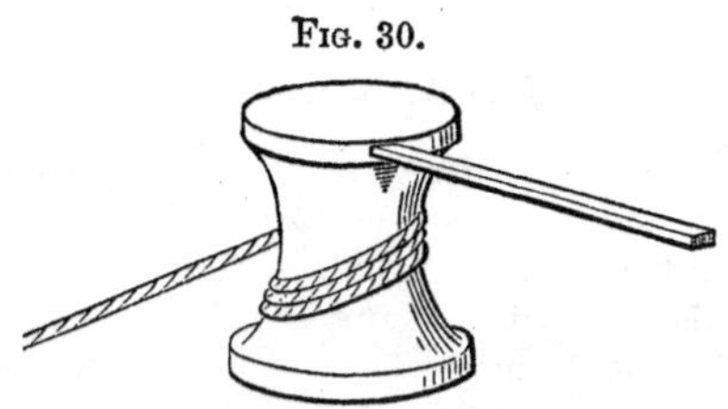

§ 93. **Toothed Wheels.**—Wheels having teeth that fit into one another do the work of levers acting continuously. The mechanical principle of these wheels is the same as that of the lever or wheel and axle.

A very simple case is what is known as the rack and pinion. It consists of a small wheel, with cogs

Fig. 31.

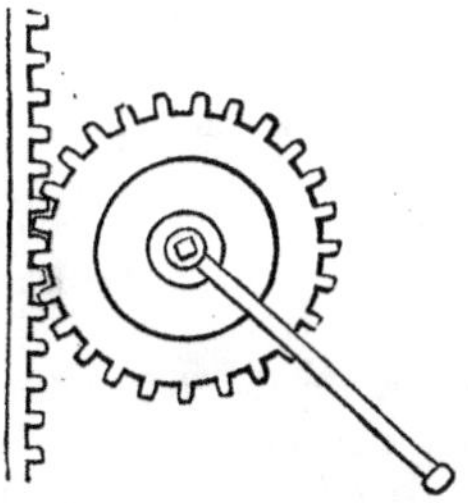

as teeth, that is made to work into a vertical bar likewise filled with teeth. In this way motion round an axis is converted into motion in a straight line. The piston of a double-barrelled air-pump is worked by such a contrivance.

EXERCISES.

1. What is the diameter of a wheel if a power of 3 ozs. is just able to move a weight 12 ozs. that hangs from the axle, the radius of the axle being 2 ins.?
2. What is the mechanical advantage of a wheel and axle in which the diameter of the axle is 3 ins., and the radius of the wheel 12 ins.?
3. Is the mechanical advantage of wheel and axle increased or diminished by lessening the radii of wheel and axle by the same amount?
4. If a weight of 20 ozs. be supported on wheel and axle by a force of 4 ozs., and the radius of the axle is $\frac{2}{3}$ in., find radius of wheel.
5. A capstan is worked by a man pushing at the end of a pole. He exerts a force of 50 lbs., and walks 10 ft. round for every 2 ft. of rope pulled in. What is the resistance overcome?
6. A man whose weight is 140 lbs. is just able to support a weight that hangs over an axle of 6 ins. radius, by hanging to the rope that passes over the corresponding wheel, the diameter of which is 4 ft.; find the weight supported.
7. If the radii of wheel and axle be as 10 is to 4, and weights of 3 ozs. and 8 ozs. hang from them, which will descend?
8. If the difference between the diameter of a wheel and the diameter of the axle be six times the radius of the axle, find the greatest weight that can be sustained by a force of 60 lbs.
9. If the radius of the wheel be n times as great as that of the axle, and t be the maximum tension of the string on the wheel, find the greatest weight that can be raised.
10. If the radius of the wheel is three times that of the axle, and the string round the wheel can support a weight of 40 lbs. only, find the greatest weight that can be lifted.

XV. *The Pulley.*

§ 94. The pulley consists of a circular plate or disc, the circumference of which is generally grooved to receive a cord which passes over it. The pulley is made to revolve freely about an axis, fixed into a

Fig. 32.

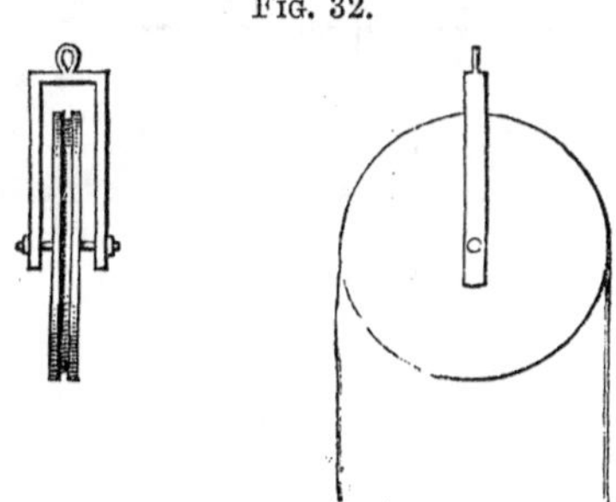

framework called the block. When the axis of the pulley is fixed the pulley is called a *fixed* pulley; but where it can ascend or descend it is called a *moveable* pulley.

§ 95. **Fixed Pulley.**—The fixed pulley can only change the direction of a force. Wherever it is required to change a pushing into a pulling force the fixed pulley can be advantageously employed. It works, however, without any mechanical advantage. The mechanical principle involved in all calculations with respect to the pulley is the constancy of the force of tension in all parts of the same string (§ 60). It should be observed that in the pulley, as in all the mechanical powers, advantage is taken of the forces that are latent in the molecular structure of beams, cords, &c. Thus, the pulley requires to be attached to a beam, and this beam must be strong enough to

Fig. 33.

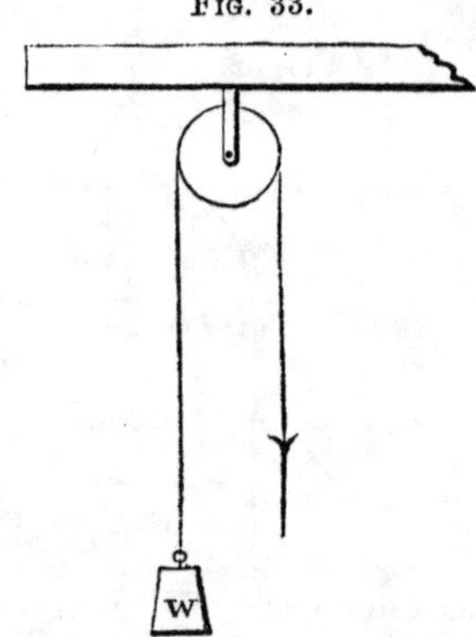

support the force exerted by the weight and the power employed to raise it. In the pulley the string is supposed to be perfectly flexible, and the friction is neglected.

§ 96. **The Single Moveable Pulley.**—A cord is fixed to a beam at A and passes over the moveable pulley $B\,D$. The other end of the cord may be at once supported, or may pass over a fixed pulley C, and bear the weight P. Now, if the machine be in motion it is clear that for every inch W rises P will descend two inches, and consequently the equation of work gives us $W = 2P$ or $P = \frac{W}{2}$. The same result may be arrived at by supposing the machine at rest and P supporting W. In this case, since the

FIG. 34.

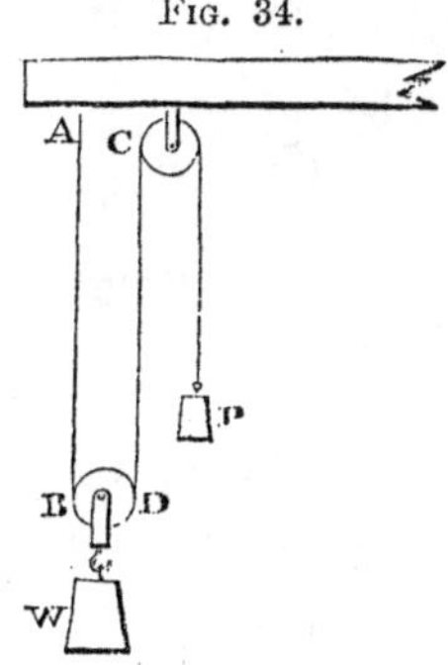

tension in the same string is always the same, and the weight W is supported by the two strings $A\,B$, $D\,C$, in each of which the tension is P, we have $W = 2P$ or $\frac{W}{P} = 2$.

The mechanical advantage with a single moveable pulley equals 2.

The single moveable pulley may be used with other combinations. Thus, if A and B (fig. 35) be two fixed pulleys separated by an interval equal to the diameter of either, and if C be a pulley of twice the diameter of A or B, and if one end of the cord be fixed to C whilst the other, after passing over the three pulleys, supports a weight P, then, whether the machine be in uniform motion or at rest, $W = 3P$.

Fig. 35.

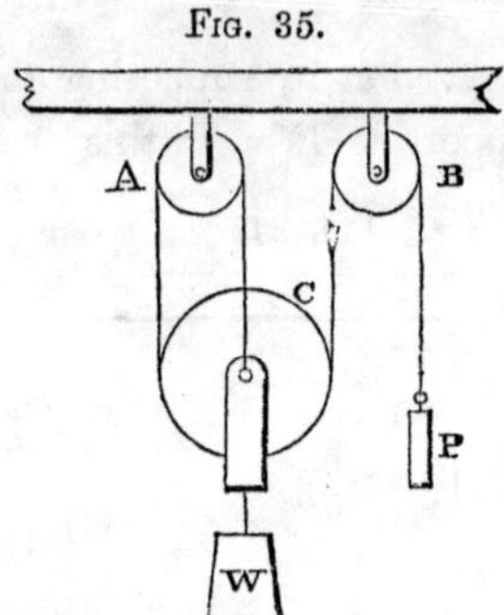

If a man be suspended from a moveable pulley and support himself by holding on to the other end

Fig. 36.

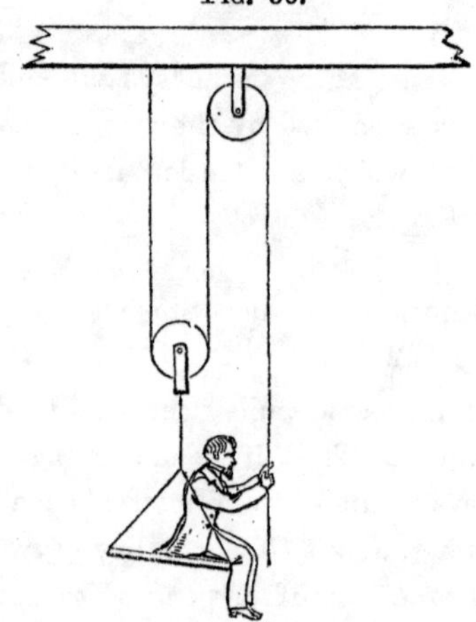

of the string, his real weight is diminished by the force with which he pulls, so that if W be his actual weight, his pressure on the pulley to which he is attached is $W - P$, and since this weight is divided between the two strings supporting the weight, $W - P = 2P$ or $W = 3P$. That is, he pulls with a force equal to one-third of his weight.

§ 97. **Combinations. First system, in which each Pulley hangs by a separate string.**—Each string has one end attached to a fixed point in the

FIG. 37.

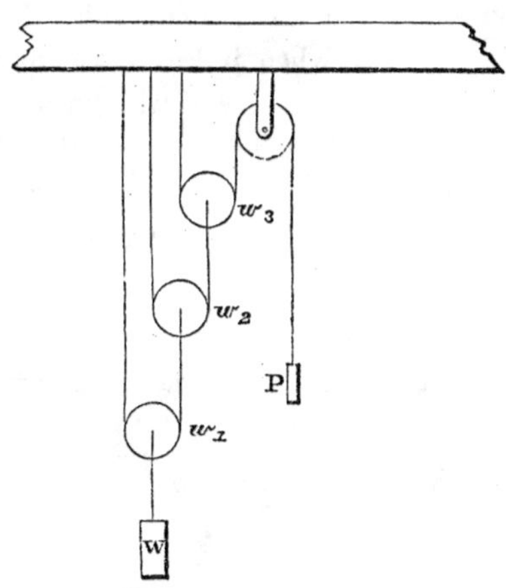

beam, and all, except the first, have the other end attached to a moveable pulley. The diagram explains the arrangement.

It is clear that if the lowest pulley to which W

is attached ascend 1 inch, the next pulley rises 2 inches, the next 4 inches, and P descends 8 inches. Thus, with three moveable pulleys the equation of work gives us $W = 8\,P = 2^3\,P$. If there be n *moveable* pulleys $W = 2\ P$ or $\frac{W}{P} = 2^n$.

Applying the principle of tension, we see that the tension in the first string is P throughout, in the string immediately beneath this it is $2\,P$ throughout, in the next $4\,P$, and the double tension of $4\,P$ supports W. In other words, $W = 8\,P$.

The pressure on the beam in this case is $W + P = 9\,P$, and if the cord when it leaves the highest pulley be at once supported by P without passing over a fixed pulley, we see that the weight W is jointly supported by the beam and the power; and the more the strength of the beam is utilised the less is the magnitude of P.

§ 98. We will now consider what alterations must be made if the weights of the moveable pulleys be taken into account.

Let w_1 be the weight of the lowest pulley.

Let w_2 " " next " and so on.

Then $W + w_1$ is the weight supported by the lowest cord, and the tension in each part of it is $\frac{W + w_1}{2}$. Similarly the tension in each part of the cord immediately above it is $\frac{1}{2}\left(\frac{W+w_1}{2}+w_2\right)$ $=\frac{W}{4}+\frac{w_1}{4}+\frac{w_2}{2}$. The tension in each part of the string above that is

$$\frac{1}{2}\left(\frac{W}{4}+\frac{w_1}{4}+\frac{w_2}{2}\right)+\frac{w_3}{2}=\frac{W}{8}+\frac{w_1}{8}+\frac{w_2}{4}+\frac{w_3}{2}.$$

Now, the last tension equals P. If therefore we have three moveable pulleys, the weights of which are w_1, w_2, w_3,

$$P=\frac{W}{2^3}+\frac{w_1}{2^3}+\frac{w_2}{2^2}+\frac{w_3}{2}.$$

Example.—With three moveable pulleys, arranged as above, each of which weighs 8 ozs., what weight can be supported by a force of 1 lb.?

Here $$P=\frac{W}{8}+\frac{w}{8}+\frac{w}{4}+\frac{w}{2}=\frac{W}{8}+7$$

$$\therefore 16 \text{ ozs.}=\left(\frac{W}{8}+7\right)\text{ozs.}$$

$$\therefore W=72 \text{ ozs.}$$

§ 99. **Second System, in which the same string passes round all the Pulleys.**—Suppose there are three pulleys in each block, then it is clear that if W ascend 1 inch, P descends 6 inches, or $W=6P$. Since the tension is the same in every part of the string, and is everywhere equal to P, and since we have six strings supporting W, it is clear, supposing all the strings to be parallel, that $W = 6\,P$.

If we have n pulleys in both blocks and w is the weight of the lower block, $W + w = n\,P$. The

Fig. 38. Fig. 39.

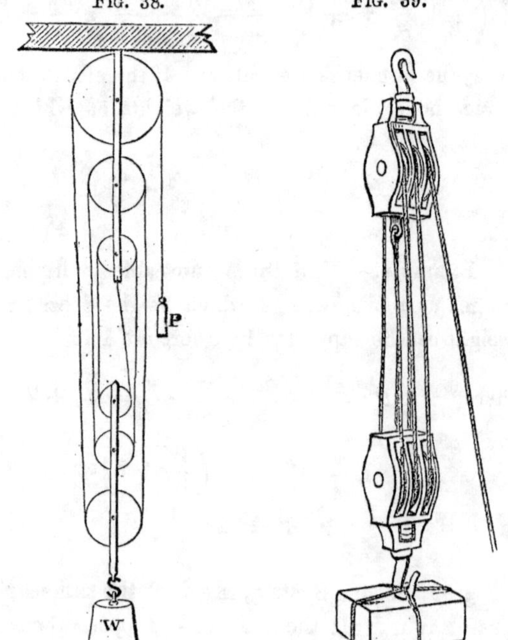

pressure on the beam is $W + P +$ weights of both blocks. This system is most commonly employed on account of its superior portability. The several pulleys are generally mounted on a common axis and enclosed in a single block, as shown in fig. 39.

§ 100. **Third System, in which each string is attached to the weight.**—In this system one end of each string is attached to the bar from which the

FIG. 40.

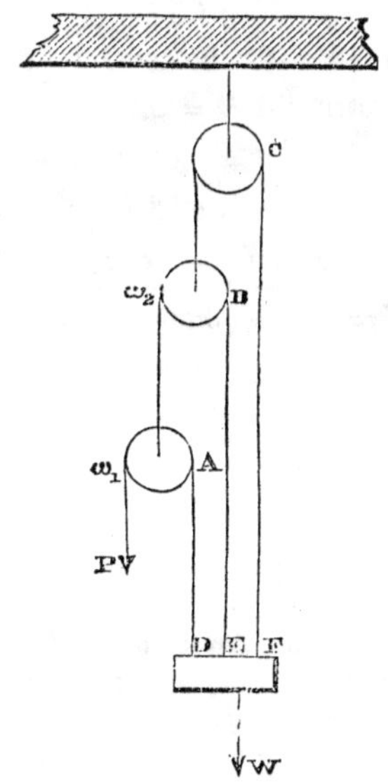

weight hangs, and the other supports a pulley. The power P acts at the unattached end of the string $A\,D$. The weight W is supported by the tensions in the three strings $A\,D$, $B\,E$, and $C\,F$.

These tensions are respectively equal to P, $2P$, and $4P$ $\therefore W = P + 2P + 4P = 7P$. If there be n pulleys

$$W = P + 2P + 2^2 P + 2^3 P + \cdots + 2^{n-1} P,$$

the sum of which series is shown, in books on algebra, to be equal to $(2^n - 1) P$, $\therefore \frac{W}{P} = 2^n - 1$.

If the weights of the pulleys be considered, we have

tension in the string $AD = P$

" " " $BE = 2P + w_1$

" " " $CF = 4P + 2w_1 + w_2$

and the sum of all these tensions equals W

$$\therefore W = 7P + 3w_1 + w_2.$$

In this system the weights of the *moveable pulleys assist* P; in the two former systems they act against it.

EXERCISES.

1. A man weighing 140 lbs. forces up a weight of 80 lbs. by means of a fixed pulley under which he stands; find his pressure on the floor.
2. Find the power which will support a weight of 600 lbs., with three moveable pulleys arranged as in the first system.
3. What force is necessary to raise a weight of 120 lbs. by an arrangement of six pulleys in which the same string passes round each pulley?
4. If, in the preceding problem, the weight of the block be 6 lbs., what additional force will be required?

5. Find the mechanical advantage of a system of three moveable pulleys arranged as in the annexed diagram.

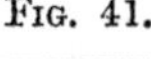

Fig. 41.

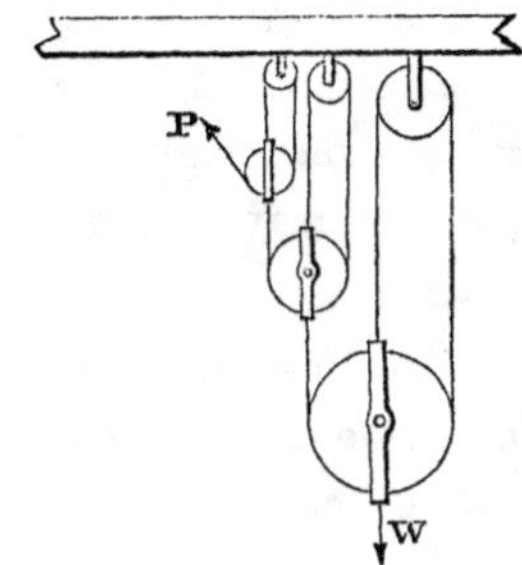

6. If three moveable pulleys, the weights of which are 2 ozs., 4 ozs., and 8 ozs., be arranged as in the first system, what is the least force that will raise a weight of 104 lbs.?
7. If there be equilibrium between P and W with three pulleys in that system in which each string is attached to the weight, what additional weight can be raised if 2 lbs. be added to P?
8. A man weighing 150 lbs. raises a weight of 4 cwt. by a system of four moveable pulleys arranged according to the first system; what is his pressure on the ground?
9. What would be the difference in his pressure if each pulley weighed 4 ozs.?
10. Find the power necessary to sustain a weight of 100 lbs. with three moveable pulleys arranged according to the third system, the weights of the pulleys being 8 ozs., 6 ozs., and 4 ozs. respectively.

XVI. *The Inclined Plane—Wedge—Screw.*

§ 101. By means of an inclined plane, a body can be raised to a certain height by the application of a power less than the weight of the body. All roads that are not level may be regarded as inclined planes.

If $A\,B\,C$ be an inclined plane, then $A\,B$ is called the *length* of the plane, $B\,C$ the *base* of the plane, and $A\,C$ the *height* of the plane.

§ 102. **Uniform motion on an Inclined Plane by application of a force parallel to the Plane.**— This case has already been partly considered. If a

Fig. 42.

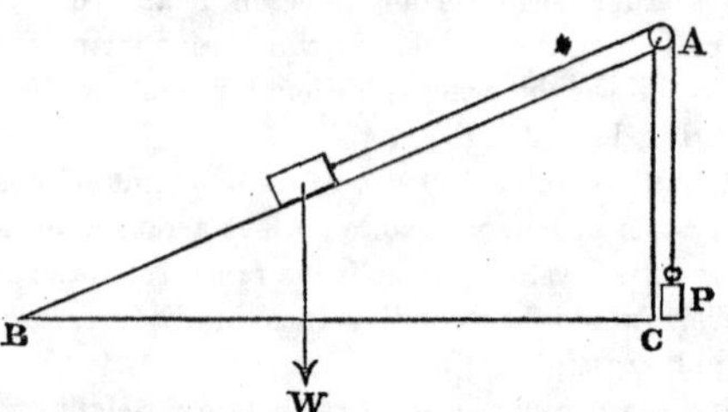

body the weight of which is W move from B to A by the continued action of a force P, then the work done by P is $P \times A\,B$, and the work done by W against gravity, *i.e.* in the direction $A\,C$, is $W \times A\,C$, since W has been raised through $A\,C$;

$$\therefore P \times A\,B = W \times A\,C, \text{ or } \frac{W}{P} = \frac{l}{h}.$$

§ 103. Uniform motion on Inclined Plane by

FIG. 43.

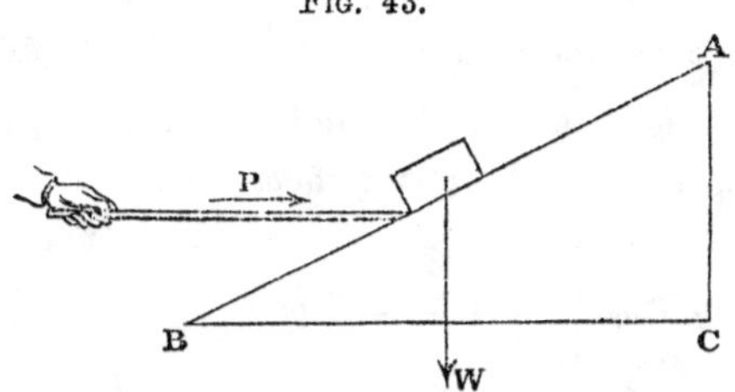

application of horizontal or pushing force.—If the force P act in the direction $B\,C$, the work done by P is $P \times B\,C$,

$$\therefore P \times B\,C = W \times A\,C, \text{ or } \frac{W}{P} = \frac{B\,C}{A\,C} = \frac{b}{h}.$$

If a weight W be just supported by two forces one of which P is parallel to the length and the other Q acts in a direction parallel to the base of the plane, then since P alone can support a weight equal to $\frac{l}{h} P$, and Q alone a weight equal to $\frac{b}{h} Q$; if W be the weight supported by both,

$$W = \frac{l}{h} P + \frac{b}{h} Q$$

or

$$W h = P\,l + Q\,b.$$

§ 104. **Example.**—Two inclined planes of different lengths, but having a common height, are placed back to back, and two weights P and Q, connected by a string that passes over a pulley at their summit, are moving uniformly on them; if l and l' be the lengths of the planes find the ratio of P to Q.

If the weights are moving uniformly, it is clear that the tension in the string must equal the force just sufficient to support either. Call T the tension of the string, then $T = \frac{h}{l} P = \frac{h}{l'} Q$

$$\therefore \frac{P}{l} = \frac{Q}{l'}$$

or $$P : Q :: l : l'.$$

The Wedge.

§ 105. **The Wedge** is a double inclined plane, moveable instead of fixed, as in the cases considered, and used for separating bodies. The force is applied in a direction perpendicular to the height of the plane, *i.e.* parallel to the base, and the resistance to be overcome consists of the molecular attractions of the particles of the body which are being separated. This resistance acts in a direction at right angles to the inclined surface of the wedge, or length of the plane. The line $A B$ is called the back of the wedge.

Suppose the wedge has been driven into the material a distance equal to $D C$ by a force P acting in the direction DC, then it is clear that the work done by P is $P \times DC$. Draw DE, DF perpendicular to AC, BC. Then, since the points E and F were originally together, the work done against the resistance R is $R \times DE + R \times DF = 2R \times DE$. Hence the equation of work gives

$$P \times DC = R \times 2DE \therefore \frac{R}{P} = \frac{DC}{2DE}.$$

But $\frac{DC}{DE} = \frac{AC}{AD}$ by similarity of triangles

$$\therefore \frac{R}{P} = \frac{AC}{2AD} = \frac{AC}{AB} = \frac{\text{length of one of the equal sides}}{\text{back of the wedge}}$$

Fig. 44.

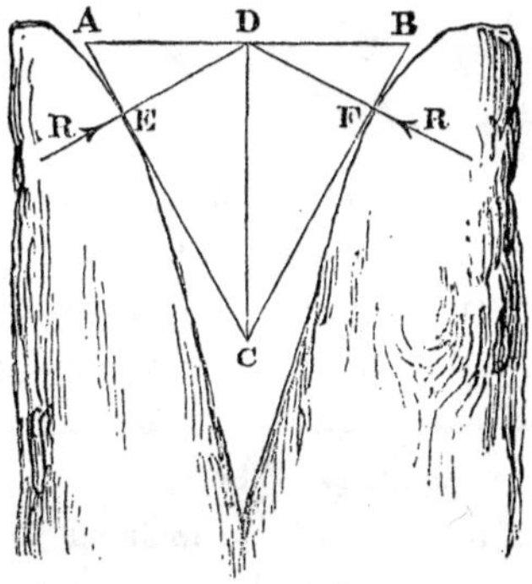

This shows that as the size of the back of the wedge is lessened the mechanical advantage of the wedge is increased. Knives, choppers, chisels and many other implements are examples of the wedge.

§ 106. In the action of the wedge a great part

of the energy of the power is employed in cleaving the material into which it is driven. The force required to effect this is so great that instead of applying a continuous pushing force perpendicular to the back, a series of blows is generally imparted to it. In this way a large amount of energy is directed for an indefinitely short period of time against the molecular attractions of the body. When the resistance to be overcome is very great a series of *impulsive forces* is more effectual than a continuous moving force. A certain amount of energy is stored up in a descending hammer which is at once set free when the blow is given.

The Screw.

§ 107. **The Screw** is a machine which is sometimes employed to overcome resistance and sometimes to multiply pressure. It consists of a cylinder with a uniform projecting thread traced round its surface and inclined at a constant angle to lines parallel to the axis of the cylinder. The thread of the screw may be formed by wrapping an inclined plane round the cylinder, the base of the plane corresponding with the circumference of the cylinder, and the height of the plane with the distance between the thread. The threads are of different shapes: they may be square or triangular. The power is usually transmitted by connecting the screw with a concave cylinder, called the nut, having a spiral cavity on its surface, corresponding to the spiral projections.

Fig. 45.

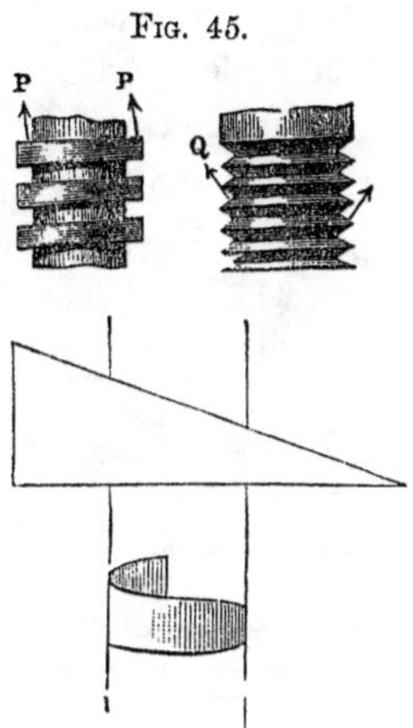

The power is almost always applied at the end of a lever fixed to the centre of the cylinder; but

Fig. 46.

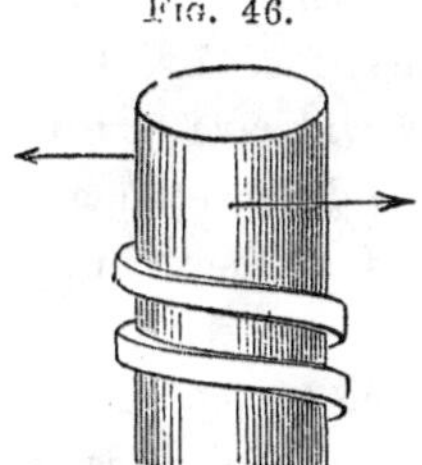

we may suppose it to act at the surface of the cylinder and in the direction of the tangent at any point. It is evident that a screw never requires any pressure in the direction of its axis. The cylinder must be made to revolve only ; and this can be effected by a force acting at right angles to the extremities of its diameter, or of its diameter produced.

If the power P acts at A at right angles to $A\,B$, and the cylinder perform one revolution, the power

Fig. 47.

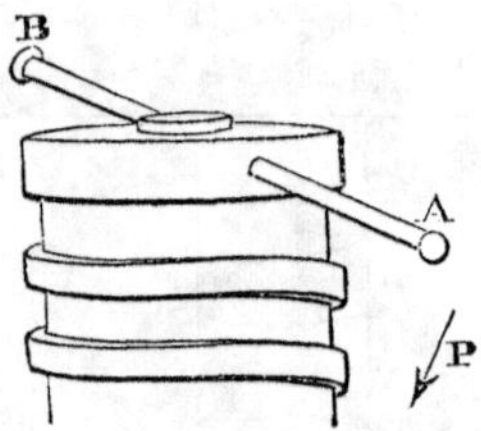

P will have acted through a distance equal to the circumference of the circle, of which $A\,B$ is the diameter; and the work done by P will be $P \times$ circumference of circle described by $A\,B$. During the same time, the screw will have moved in the direction of its axis through a distance equal to the distance between two threads of the screw, and this is the direction in which the resistance is encountered. Hence the work done against the resistance W is $W \times$ distance between the threads.

If C be the circumference described by the extremity of P's arm and d be the distance between two adjoining threads, and if P and W be respectively the force applied and the pressure produced,

$$P \times C = W \times d, \text{ or } \frac{W}{P} = \frac{C}{d}.$$

EXERCISES.

1. What weight can be supported on a plane by a horizontal force of 10 ozs., if the ratio of height to the base is $\frac{3}{4}$?
2. With a force of 3 ozs. acting parallel to the plane, what weight can be supported on a plane that rises 2 in 7?
3. A plane rises 3 in 8; what force parallel to it is required just to move a weight of 10 ozs.?

   A plane rises 2 in 7; what force parallel to it is required just to move a weight of 12 ozs.?

   A plane rises 5 in 9; what force parallel to it is required just to move a weight of 16 ozs.?

   A plane rises $1\frac{1}{4}$ in $12\frac{1}{2}$; what force parallel to it is required just to move a weight of 20 ozs.?
4. Find the inclination of the plane if a horizontal force of 5 ozs. can just move a weight of 12 ozs.
5. The inclination of a plane is 30°, and a weight of 10 ozs. is supported on it by a string bearing a weight at its extremity, which passes over a smooth pulley at its summit; find the tension in the string.
6. Find the horizontal force necessary to support a weight of 1 lb. on a plane that rises 3 in 5.
7. The angle of a plane is 45°; what weight can be supported by a horizontal force of 3 ozs. and a force of 4 ozs. parallel to the plane, both acting together?
8. Two planes, having the same height, are placed back to back, and two weights of 7 ozs. and 10 ozs. connected by a string passing over the summit move uniformly upon them; find the ratio of the lengths of the planes.
9. A body weighing 7 ozs. is supported on a plane that

rises 1 in 7 by a force that acts parallel to the plane; if 3 ozs. be added to the weight, what force will be required to support it?

10. A heavy body is just able to be pulled up a plane by a force parallel to it and equal to $\frac{1}{5}$ of the weight of the body; find the ratio of the height to the base of the plane.

11. Two weights hang over a pulley fixed to the summit of a smooth inclined plane, on which one weight is supported, and for every 3 ins. that the one is made to descend the other rises 2 ins.; find the ratio of the weights and the length of the plane, the height being 18 ins.

12. In a screw which has seven threads to the inch, find the pressure that can be produced by a force of 6 lbs. applied at the circumference, the radius of the cylinder being 1 in.

13. If the circumference of a screw be 10 ins., what force must be applied to overcome a resistance of 30 lbs., the distance between the threads being $\frac{1}{4}$ in.?

14. How many turns must be given to a screw formed upon a cylinder whose length is 10 ins., and circumference 5 ins., that a power of 2 ozs. may overcome a pressure of 100 ozs.?

15. A screw is made to revolve by a force of 2 lbs. applied at the end of a lever 3·5 ft. long; if the distance between the threads be $\frac{1}{2}$ in., what pressure can be produced?

16. If a power of 1 lb. describe a revolution of 3 ft. whilst the screw moves through $\frac{1}{4}$ in., what pressure will be produced?

17. The circumference described by the power is 4 ft., and the distance between the threads is $\frac{1}{3}$ in.; what power is required to produce a pressure of 1 ton?

EXAMINATION.

1. What is the function of a machine? What is meant by the *modulus* of a machine?
2. Apply the law of energy to the work done by a machine against the resistance and against friction.
3. Give instances of the three kinds of levers. Can the first order be worked at a mechanical disadvantage?
4. Friction has been said to be 'not the destroyer but the converter of energy.' Explain this statement.
5. Explain the action of a wedge. Give examples of its use.
6. In the wheel and axle is there any advantage in having the rope that passes round the wheel thicker than that which passes round the axle?
7. The radius of the wheel being three times that of the axle, and the string on the wheel being only strong enough to support a tension equivalent to 30 lbs., find the greatest weight which can be lifted.
8. If a weight W be supported on an inclined plane by a force $\frac{W}{2}$ parallel to the plane, what is the inclination of the plane?
9. The ratio of the height of a plane to its length is 2 : 15; what horizontal force is necessary to support a weight of 10 lbs.?
10. The inclination of a plane is 30°; a weight of 80 lbs. being placed upon it, a force of 63 lbs. is required to pull the weight up the plane; what is the coefficient of friction?
11. Two inclined planes, of the same height, one of which is 8 ft. long, and the other 5 ft., are placed so as to slope in opposite directions and so that their summits coincide. A weight of 20 ozs. rests on the shorter plane, and is connected by a string passing over a

pulley at the common summit of the two planes with a weight resting upon the longer plane; how great must this weight be to prevent motion?

12. Find the relation between P and W in a system of pulleys arranged as in the annexed diagram, supposing the

Fig. 48.

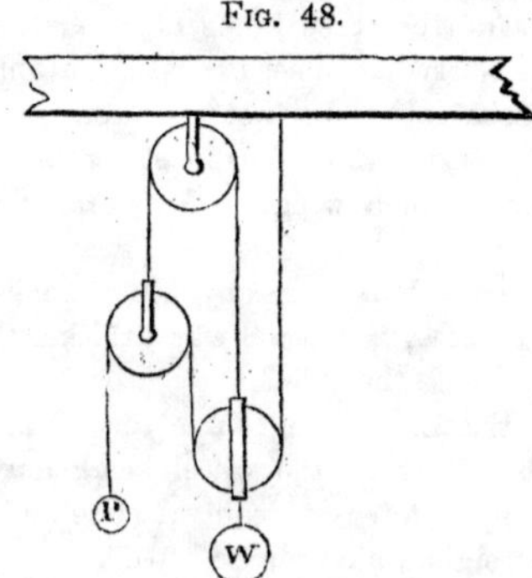

weight of each of the moveable pulleys to be w.

13. Neglecting the weights of the pulleys, what pressure would a man, whose weight is 160 lbs., exert on the ground in raising a weight of 500 lbs. by means of the above combination?

14. Describe the screw and point out its relation to the inclined plane.

15. What is the mechanical advantage of a screw?—*Matriculation*, 1869.

16. In a system of one fixed, and four moveable pulleys, in which one end of each string is fixed to a beam, find the relation between the power and the weight (neglecting the weights of the pulleys), when one of the strings is nailed to the pulley round which it passes.

What is the force exerted on the beam to which the strings are attached?—*Ib.*, 1869.

17. Find the relation between the power and the weight in the third system of pulleys (in which each string is attached to the weight), when the weights of the pulleys are disregarded. Also show what the relation would be if the weights of the pulleys were taken into account.—*Matriculation*, Jan. 1871.

18. A wheel-and-axle is used to raise a bucket from a well. The radius of the wheel is 15 ins., and while it makes seven revolutions, the bucket, which weighs 30 lbs., rises $5\frac{1}{2}$ ft. Show what is the smallest force that can be employed to turn the wheel. Upon what general principle is your answer founded?—*Ib.*, Jan. 1872.

19. Ten weights, each of 20 lbs., are to be lifted to a height of 8 ft. from the ground. Show how a system of pulleys might be arranged so that, disregarding friction and the weight of the pulleys, all the weights could be lifted together by exerting a force equal to one of them. Show that the distance through which this force would have to act would be the same as when the weights were raised one by one by the same power.—*Ib.*, Jan. 1873.

20. A man sitting upon a board suspended from a single moveable pulley pulls downwards at one end of a rope which passes under the moveable pulley and over a pulley fixed to a beam overhead, the other end of the rope being fixed to the same beam. What is the smallest proportion of his whole weight with which the man must pull in order to raise himself?—*Prelim. Scientific M.B. Exam.*, 1869.

21. With what force would he require to pull upwards, if the rope, before coming to his hand, passed under a pulley fixed to the ground, as well as round the other two pulleys?—*Ib.*, 1869.

22. In the third system of pulleys, in which each string is attached to the weight, each pulley weighs $3\frac{1}{2}$ ozs. Find the weight which will be supported by the pulleys alone, when there are five moveable pulleys.—*Ib.*, 1872.

23. In the third system of pulleys, in which each string is attached to the weight, show that the power multiplied by its descent equals the weight multiplied by its ascent, neglecting the weights of the pulleys and all friction.—*Prelim. Scient. Exam.*, 1873.

24. Suppose that we have four weightless pulleys, three moveable and one fixed, forming an example of the first system, and that the weight is a man weighing 160 lbs., find what pull the man must exert on the power end of the rope in order to raise himself thereby.—*Ib.*, 1874.

STATICS—REST.

CHAPTER VII.

THEORY OF EQUILIBRIUM.

XVII. *General Considerations—Forces in the same Straight Line.*

§ 108. **Problem of Statics.**—The problem of statics is to determine the conditions under which several forces acting on a body produce equilibrium. To solve this problem, it is often convenient to find the single force, if one exists, that will produce the same effect as the other forces taken together. This single force, which can replace several other forces, is called the *resultant*, and the forces themselves are called *components*. The resultant being determined, the problem of statics is solved when the force is found which will keep this resultant in equilibrium. We shall be occupied, therefore, for some time in finding the resultant of forces acting in different directions.

§ 109. Method of estimating and representing Statical Forces.—Forces in statics are supposed to be prevented by some kind of resistance from producing motion. They are generally measured by the *pressure* they are capable of producing as reckoned in weight. The knowledge of a force implies the knowledge of (1) its point of application, (2) its direction, (3) its magnitude.

All these elements can be represented geometrically. A straight line can be drawn from any point and in any direction; and if a unit of length represent a unit of force, the number of units of length in the line will represent the number of units of force. Or, two forces P and Q will be represented by two lines, AB and CD, when $P : Q :: AB : CD$. Care should be taken to distinguish the force AB from the force BA, since these two forces, though equal in magnitude, act in opposite directions.

§ 110. Forces in the same Straight Line.—If two forces act upon a body it is clear that in order that they should produce no effect they must act (1) at the same point, (2) in opposite directions, and (3) they must be equal in magnitude.

If instead of two we have several forces acting at a point and in the same straight line, it is evident, that for equilibrium the tendency to motion in one direction must be counterbalanced by the tendency to motion in the opposite direction, or the sum of the forces in one direction must equal the sum of the forces in the opposite direction; *i.e.* the *algebraical* sum of the forces must vanish.

If X_1, X_2, X_3...be the several forces and $\Sigma(X)$ denote their algebraical sum, then the condition of equilibrium is $\Sigma(X) = O$.

§ 111. **Forces in Equilibrium acting at a Point.**—Since the resultant is that force which produces the same effect as all the other forces taken together, it is evident that when the forces produce equilibrium their joint effect equals zero, or the resultant vanishes. The resultant being determined, the conditions of equilibrium are the conditions that must hold good in order that this resultant should become zero. The solution of the equation $R = O$, when R is the numerical value of the resultant, will always determine the conditions of equilibrium when several forces act at a point.

If any number of forces acting at a point be in equilibrium, and one of them be removed, the resultant of all the rest is equal in magnitude, but opposite in direction to the removed force; for, since the forces were originally in equilibrium, the removal of one force must destroy the equilibrium, since all the other forces served to counteract the effect of this one. But the single force which will counteract the effect of another force is one equal in magnitude and opposite in direction, and therefore a force equal in magnitude to the removed force, but opposite in direction, produces the same effect as all the remaining forces, or, the removed force reversed is the resultant of the rest.

§ 112. It will be seen that when forces are in equilibrium motion may be produced either by the removal of one of the forces or the application of a new one. A balloon resting in mid-air will begin to rise if some ballast be thrown out, or to sink if gas be allowed to escape. A body in motion may be acted upon by one force only; but to preserve a body in equilibrium two forces at least must cooperate.

§ 113. The resultant of any number of forces should be carefully distinguished from the force that keeps them in equilibrium; for these two forces though equal in magnitude are opposite in direction.

EXERCISES.

1. A body rests on a perfectly smooth table, and to one end of it is tied a string which is stretched by weights of 3 ozs., 5 ozs., and 7 ozs.; and to the other end a string that is stretched by weights of 1 oz., 8 ozs., and 2 ozs.; what additional weight is necessary to preserve equilibrium?
2. A string AD is suspended at A; at B, a point in the same string, a weight of 3 ozs. is attached, at C a weight of 4 ozs., at D a weight of 5 ozs.; find the tension in each part of the string.

3. If a force of 13 lbs. be represented by a line of $6\frac{1}{2}$ ins., what line would represent a force of $7\frac{1}{2}$ lbs.?
4. The diagonal of an oblong is 5 ins., and one of its sides is 3 ins. Two forces acting in the same straight line are to one another as the sides of the oblong; what length of line would represent the force necessary to produce equilibrium?

XVIII. *Composition of Forces acting at a Point, but not in the same Straight Line.*

§ 114. We will, first, consider the case of *two* forces acting at a point but not in the same straight line.

It has been seen in Lesson II. that if *O A*, *O B* represent two velocities with which a body tends to move, then *O C*, the diagonal of the parallelogram formed about *O A* and *O B*, represents the resultant velocity, and *C* is the point which the body would reach in the time occupied in passing from *O* to *A* or from *O* to *B*. What is true of velocities is true of the forces that would produce them, and if *O A* and *O B* represent two forces, *O C* represents the force equivalent to both, and *C O* the force that would keep them in equilibrium. This proposition, known as the Parallelogram of Forces, follows at once from Newton's second law of motion, and may be enunciated thus:—

If two forces acting at a point be represented in

magnitude and direction by the sides of a parallelogram, the resultant of these two forces will be represented in magnitude and direction by the diagonal of the parallelogram passing through this point.

§ 115. This proposition may be experimentally verified by passing a fine thread over two smooth pulleys fixed to a wall, as in the adjoining figure. From the ends of the string hang two weights P and

FIG. 49.

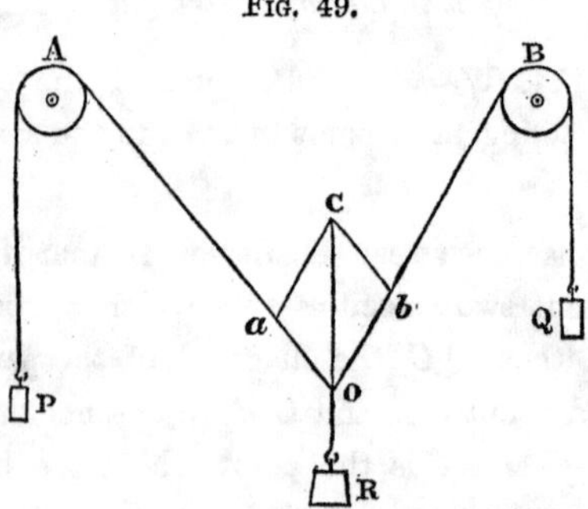

Q, and to some point in the string a third weight R is attached. When these weights are in equilibrium it will be found that a parallelogram may be constructed the sides and diagonal of which are proportional to the three weights. Thus, if P, Q and R be in equilibrium at O, and OC be taken to represent R, and Ca, Cb be drawn parallel to OB, OA, then Oa, Ob, OC will represent P, Q, and R respectively. In other words, it will be found that $Oa : OC :: P : R$, and $Ob : OC :: Q : R$.

§ 116. **Resultant of several Forces acting at the same point in different directions.**—Let P_1, P_2, P_3, and P_4 be forces acting at a point O.

Let $O A$ represent P_1
„ $O B$ „ P_2
„ $O C$ „ P_3
„ $O D$ „ P_4.

Then by the foregoing proposition the resultant of P_1 and P_2 is a force represented by $O a$; and if

FIG. 50.

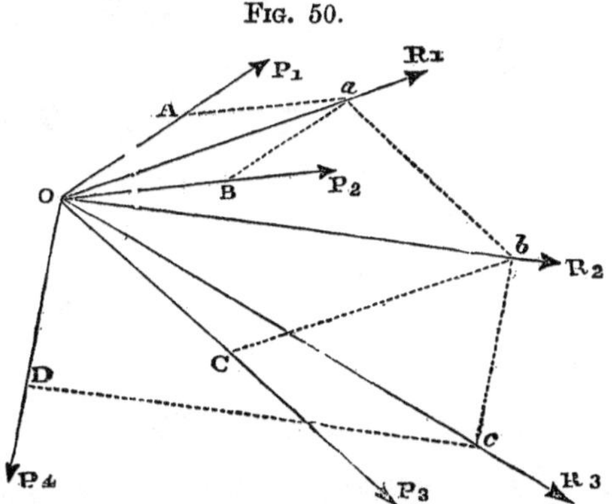

this be combined with $O C$, the new force represented by $O b$, will be the resultant of P_1, P_2, and P_3. By combining this resultant with P_4, we obtain $O c$ the resultant of P_1, P_2, P_3 and P_4; and in the same way the resultant of any number of forces may be obtained.

The above method shows how the resultant of

any number of forces may, theoretically, be obtained. In actual work, so many mathematical difficulties arise that in most cases it will be found convenient to employ another method, which will be presently considered.

§ 117. **Formula for the Resultant of two Forces acting at a point.**—Let OA, OB represent P and Q,

I. FIG. 51. II.

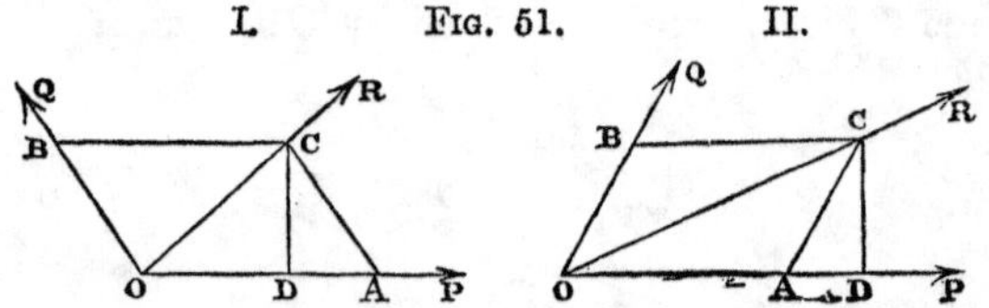

two forces acting at O. Complete the parallelogram $OABC$. Then OC represents the resultant. From C draw CD perpendicular to OA or OA produced. Then, since AC equals OB, AC represents Q. By Euclid II. 12 and 13:

$$OC^2 = OA^2 + AC^2 \pm 2\,OA \,.\, AD \quad \text{or}$$
$$R^2 = P^2 + Q^2 \pm 2P \times AD$$

The upper sign + referring to fig. II., and the lower sign − to fig. I.

Thus R can be found in terms of P and Q, whenever the value of AD can be expressed in terms of AC. This, we shall see, is in many cases possible without the aid of trigonometry.

§ 118. **Special Cases of the General Formula.**—(1) Let the two forces act at right angles. Then, since $\angle\ B\,O\,A$ is a right angle $O\,C^2 = O\,A^2 + A\,C^2$ *i.e.* $R^2 = P^2 + Q^2$, or $R = \sqrt{(P^2 + Q^2)}$.

(2) Let the angle $B\,O\,A$ be 30°. Then in fig. 51 II., the angle $C\,A\,D = 30°$ and $A\,D = A\,C\,\frac{\sqrt{3}}{2}$ or $R^2 = P^2 + Q^2 + \sqrt{3}\,P\,Q$.

(3) Let the angle $B\,O\,A$ be 45°. Then the angle $C\,A\,D = 45°$ and $A\,D = A\,C\,\frac{1}{\sqrt{2}}$

or $$R^2 = P^2 + Q^2 + \sqrt{2}\,P\,Q.$$

In the same way, by reference to fig. 51, I., the values of R^2 may be found when the angle $B\,O\,A$ is 120°, 135°, or 150°.

The foregoing results may be arranged and remembered in the following form:—

If the angle between the forces be

	30°,	$R^2 = P^2 + Q^2 + \sqrt{3}\,.\,P\,Q$
if	45°,	$R^2 = P^2 + Q^2 + \sqrt{2}\,.\,P\,Q$
„	60°,	$R^2 = P^2 + Q^2 + P\,Q$
„	90°,	$R^2 = P^2 + Q^2$
„	120°,	$R^2 = P^2 + Q^2 - P\,Q$
„	135°,	$R^2 = P^2 + Q^2 - \sqrt{2}\,.\,P\,Q$
„	150°,	$R^2 = P^2 + Q^2 - \sqrt{3}\,.\,P\,Q.$

It will be seen that the several values of the

resultant lie between two extreme values when the angle $B\,O\,A$ is 0° and 180°. In these cases $R = P + Q$ and $P - Q$ respectively. Hence R^2 is always between $P^2 + Q^2 + \surd 4 \,.\, P\,Q$, and $P^2 + Q^2 - \surd 4 \,.\, P\,Q$.

§ 119. **Examples.**—(1) Single moveable pulley with cords inclined. In considering pulleys we omitted cases in which the cords are inclined. Let C be a single moveable pulley with cords inclined at any of the above-mentioned angles. Then it is clear

FIG. 52.

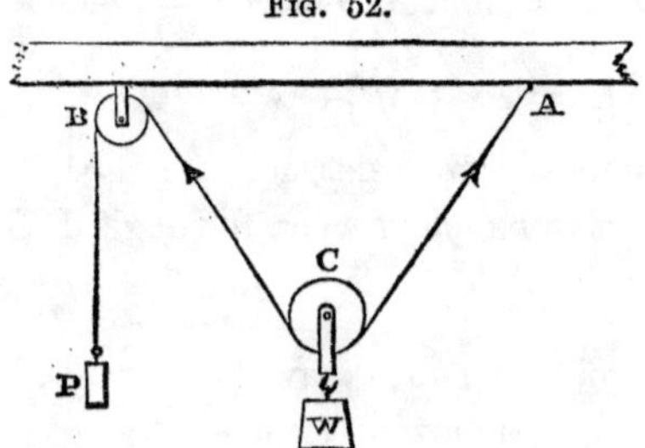

that the weight W is equal to the resultant of the tensions in the cords $C\,A$, $C\,B$, and these two tensions are equal to each other and to the power P. If the angle between the threads be 60°,

$$W^2 = P^2 + P^2 + P \times P = 3\,P^2$$

$$\therefore W = \surd 3 \,.\, P \text{ or } \frac{W}{P} = \surd 3.$$

In the same way the ratio of W to P may be found when the cords are inclined at other angles.

(2) Two equal forces act along $C B$, $B A$ two sides of the equilateral triangle $A B C$; find their resultant. The lines $C B$, $B A$ are inclined at an angle of 120°.

$\therefore R^2 = P^2 + Q^2 - P Q$, or if each of the forces equal P,

$$R^2 = P^2 + P^2 - P^2, \text{ or } R = P.$$

Hence, the resultant of two equal forces inclined at an angle of 120° is equal to either of them.

The directions in which two or more forces act are often indicated by saying that the forces act along the sides of a figure the angles of which are known. Thus two forces acting along $B C$, $B A$ act at an angle of 60°.

§ 120. **Resultant of three Forces acting at a point but not in the same plane.**—Let $O A$, $O B$, $O C$ represent P_1, P_2, P_3 acting at O. Then the

FIG. 53.

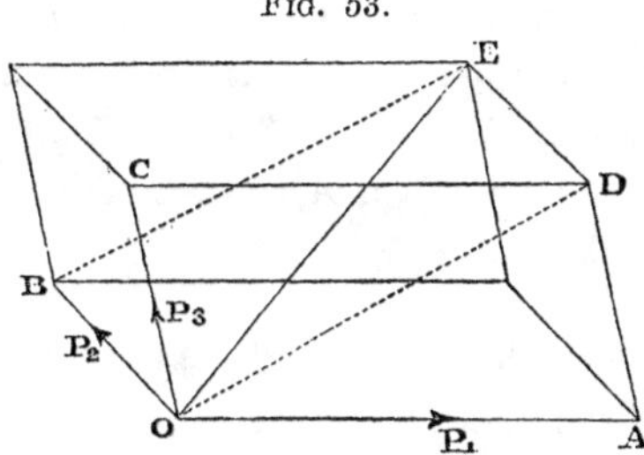

resultant of P_1 and P_3 is represented by $O D$, the diagonal of the parallelogram $O C D A$, and the resultant of this force and P_2 is represented by $O E$, the diagonal of the parallelogram $O B E D$, and of the parallelopiped of which the three lines $O A$, $O B$, $O C$ are edges.

If the three forces act at right angles to one another:

$$R^2 = P_1^2 + P_2^2 + P_3^2.$$

§ 121. **The Resultant is always nearer to the greater force.**—The line $O C$ is said to be nearer to

FIG. 54.

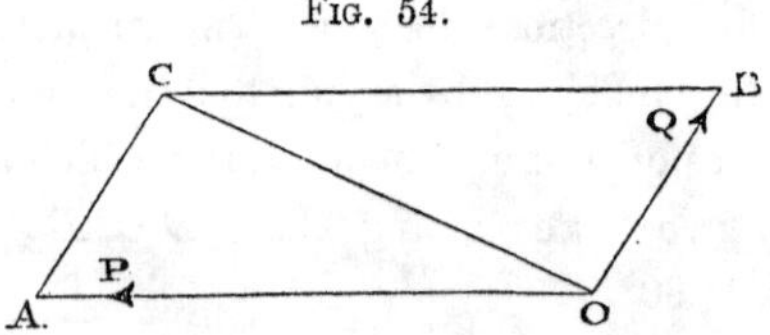

$O A$ than to $O B$, when the angle it makes with $O A$ is less than the angle it makes with $O B$. Let $O A$ and $O B$ represent two forces P and Q, of which P is the greater. Complete the parallelogram $O A C B$. Then $O A = B C$; and since $B C$ is greater than $O B$ the angle $C O B$ is greater than the angle $O C B$. But the angle $O C B = \angle C O A \therefore \angle C O B$ is greater than $\angle C O A$, and therefore $O C$ is nearer to $O A$ than to $O B$.

§ 122. **The greater the angle between two forces, the less is their resultant.**—Let P and Q be two forces and let the angle between them be, first, $A\,O\,B$, and secondly, $A'\,O\,B'$, and let the angle $A'\,O\,B'$ be greater than $A\,O\,B$. Then, since $O\,B = O\,B'$ and $B\,C = B'\,C'$, and the angle $O\,B\,C$ is greater than the angle $O\,B'\,C$, it follows that $O\,C$ is

Fig. 55.

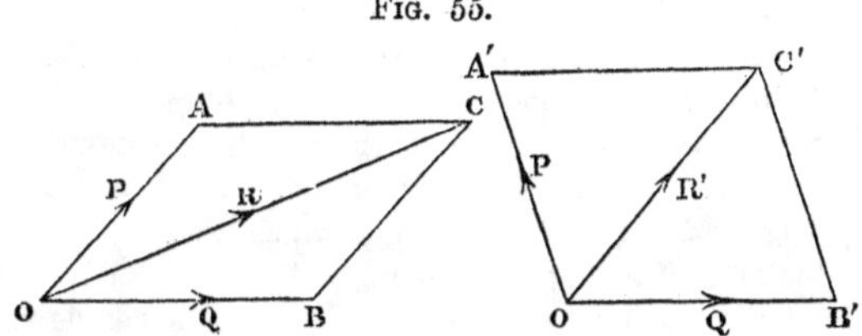

greater than $O\,C'$. Hence the resultant decreases as the angle between the forces increases.

§ 123. If two forces are represented by $A\,B$ and $A\,C$, the sides of a triangle, then their resultant will be represented by twice $A\,D$, where D bisects $B\,C$,

Fig. 56.

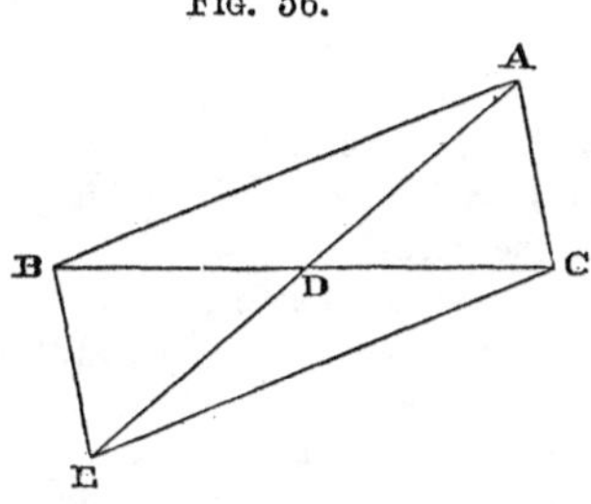

the base of the triangle. For, if the parallelogram $A\,B\,E\,C$ be completed, $A\,E$ represents the resultant of the two forces, and $A\,E = 2\,.\,A\,D$, since the diagonals of a parallelogram bisect each other. The line $A D$ is called the *median* line of the triangle $A B C$.

EXERCISES.

1. A particle is acted upon by two forces of 10 lbs. and 12 lbs., one of which is inclined at an angle of 80° to the vertical, and the other at an angle of 40° to the vertical and on the other side of it; find the magnitude of the single force which would produce the same effect as these two conjointly.
2. Find the resultant of two forces acting at an angle of 45°, one of which is twice the other.
3. Two forces, one of which is three times the other, act along the adjacent sides of a square; find the resultant.
4. Three equal forces act along the sides $A B$, $B C$, $D C$ of a square $A B C D$; find the resultant.
5. Forces of 8 lbs. and 10 lbs. act at an angle of 60°; find their resultant.
6. The resultant of two forces that act at right angles to one another is 145 lbs., and one of the forces is 144 lbs.; find the other.
7. Two forces whose magnitudes are as 3 to 4 act at a point in directions at right angles, and produce a resultant of 2 lbs.; find the forces.
8. The directions of two forces acting at a point are inclined to each other (1) at an angle of 60°, (2) at an angle of 120°, and the respective resultants are in the ratio $\sqrt{7} : \sqrt{3}$; compare the magnitude of the forces.
9. Forces of 4 lbs. and 5 lbs. act along the sides $A B$, $B C$ of an equilateral triangle; find the resultant.
10. A boat is moored in a stream by two ropes fastened to each bank, and inclined to the direction of the stream

at angles of 150° and 120° respectively. The force of the stream is equal to 500 lbs., and the tension in one of the strings is 300 lbs.; find the tension in the other.

11. Two forces of 4 ozs. and $3\sqrt{2}$ ozs. act at an angle of 45°, and a third force of $\sqrt{42}$ ozs. acts at right angles to their plane at the same point; find their resultant.

12. Three rods are joined at a point, and at right angles to one another, and a weight of $4\sqrt{3}$ lbs. hangs from their point of intersection; find the pressure transmitted through each rod.

13. Four equal forces act at a point; the first is at right angles to the second, the third is at right angles to the resultant of the first two, and the fourth is at right angles to the resultant of the other three; find the resultant of all four.

14. Three posts are placed in the ground so as to form an equilateral triangle, and an elastic ring is stretched round them, the tension of which is 6 lbs.; find the pressure on each post.

15. Six posts are fixed in the ground so as to form a regular hexagon, and a cord is passed twice round them, and pulled with a force of 100 lbs.; find the magnitude and direction of resultant pressure on each post.

XIX. *Geometrical Condition of Equilibrium when two or more Forces act at a Point.*

§ 124. **Triangle of Forces.**—If OA and OB be two forces acting at O, and the parallelogram $OABC$ be completed, OC will represent the resultant of these two forces, and OC reversed or CO will keep OA and OB in equilibrium. Since AC equals

O B, the three forces represented by *O A*, *A C*, and *C O* will be in equilibrium. Conversely, if three forces acting at a point are in equilibrium, since any one reversed equals the resultant of the others, the

Fig. 57.

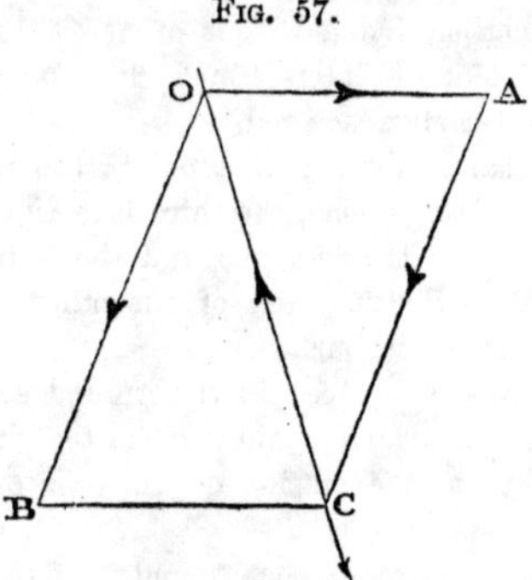

three forces can be represented by the sides of a triangle *taken in order*.

If a triangle be drawn anywhere else, with its sides respectively equal to the sides of the triangle *O A C*, it can be made to coincide with it, and will therefore represent the magnitude of the three forces in equilibrium.

The proposition known as the *triangle of forces* is generally enunciated thus:—

'If three forces acting at a point be in equilibrium, and if any triangle be drawn, the sides of which are respectively parallel or perpendicular to their directions, the forces will be to one another as the sides of the triangle; and, conversely, if the three forces are to one another as the sides of the triangle, they will be in equilibrium.'

§ 125. The proposition thus enunciated may be stated more generally, for the sides of a triangle may represent the magnitudes of the forces, though they are neither parallel nor perpendicular to their directions. Nothing more is necessary than that the triangle shall be such as is capable of being taken up and twisted into parallelism with the sides. In other words, if three forces acting at a point be in equilibrium, and there be a triangle which can be so moved that its sides shall become respectively parallel to the directions of the forces, then the sides of this

Fig. 58.

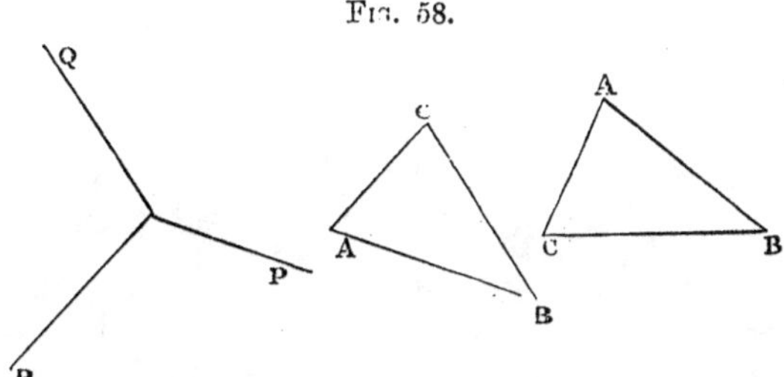

triangle, taken in order, will represent the forces; and conversely if the three forces can be represented by the sides of such a triangle, they will be in equilibrium.

Thus, if P, Q and R be three forces in equilibrium

and $A\,B\,C$ a triangle, the sides of which can be brought into parallelism with the directions of the forces, then

$$P : Q : R :: A B : B C : C A.$$

It is necessary to observe that if the sides of the triangle be not taken in order they will represent two forces and their resultant. Thus, $A\,B$, $B\,C$, and $A\,C$ represent P, Q, and their resultant R.

This proposition, viz. the triangle of forces, is the statical supplement to the parallelogram of forces, as it gives the geometrical conditions of equilibrium when three forces act at a point.

§ 126. **Examples.**—(1) Equilibrium on the inclined plane.—When a heavy body is supported on

Fig. 59.

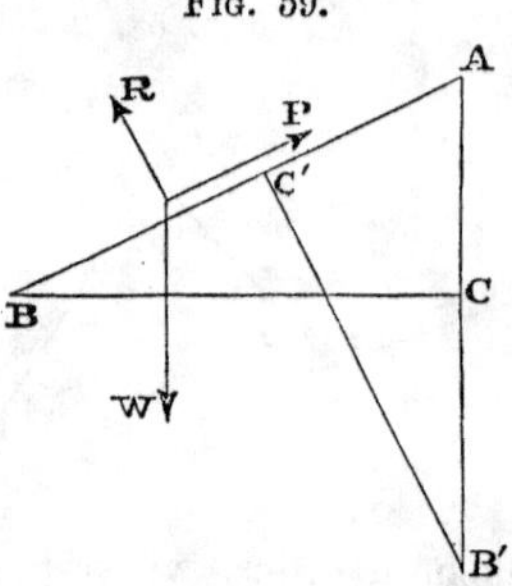

an inclined plane whether by a force parallel to the plane or parallel to the base, the relation between the forces in equilibrium can be deduced from the triangle of forces.

First. Let the force P act parallel to the plane $A\,B$. Let W be the weight of the body, and R the normal pressure or reaction of the plane. Then P, W, and R are supposed to be in equilibrium. Produce $A\,C$ to B' and make $A\,B' = A\,B$. From B' draw $B'\,C'$ perpendicular to $A\,B$. Then the triangle $A\,B'\,C'$ is in every respect equal to the triangle $A\,B\,C$, and its sides are respectively parallel to the directions of the three forces. Therefore the sides of the triangle $A\,B'\,C'$ represent the forces P, W, R. But the triangle $A\,B'\,C'$ is only a new position of the triangle $A\,B\,C$, which the original triangle can be made to assume.

$$\therefore P : W : R :: C'\,A : A\,B' : B'\,C'$$
$$:: A\,C : A\,B : B\,C$$

or $$\frac{P}{W} = \frac{A\,C}{A\,B} = \frac{h}{l} \text{ and } \frac{P}{R} = \frac{A\,C}{B\,C} = \frac{h}{b}.$$

FIG. 60.

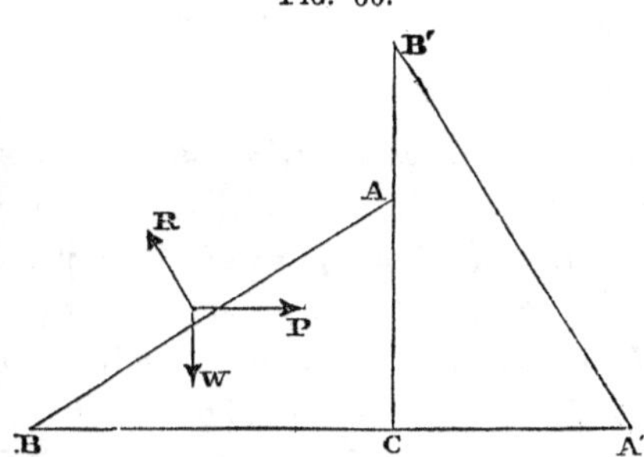

Secondly. Let the force P act parallel to the base of the plane. Then the triangle $A\,B\,C$ can be twisted into the position $A' B' C$; in which case $C A'$ is parallel to P, $A' B'$ to R, and $B' C$ to W

$$\therefore P : R : W :: C A : A B : B C$$

or $$\frac{P}{R} = \frac{C A}{A B} = \frac{h}{l} \text{ and } \frac{P}{W} = \frac{C A}{B C} = \frac{h}{b}.$$

(2) Two strings are each tied to a weight of 25 ozs. at C, and their other extremities are attached to two weights P and Q, which hang over the smooth pegs A and B, in the same horizontal line. If

Fig. 61.

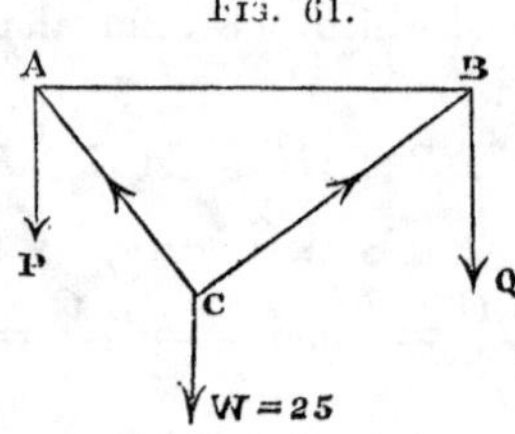

$A C$ be 3 inches, and $B C$ be 4 inches, and $A B$ be 5 inches, find P and Q.

Here the three forces are $W = 25$ ozs. and P and Q the tensions in the strings $C A$, $C B$. They act at the point C; and since $5^2 = 4^2 + 3^2$, it follows that the angle $A C B$ is a right angle (Euc. I. 48). Since $A B$, $B C$, and $C A$ are respectively perpendicular to the directions of W, of the tension in $A C$, and of the tension in $B C$, the triangle $A B C$ represents the three forces in equilibrium, and

$$W : P : Q :: AB : BC : CA$$
$$:: 5 : 4 : 3$$

$$\therefore P = \frac{4}{5} W = 20 \text{ ozs., and } Q = \frac{3}{5} W = 15 \text{ ozs.}$$

(3) Required the least horizontal force necessary to draw a heavy wheel over an obstacle the height of which is h, situated on the horizontal plane on which the wheel rests.

Let W be the weight of the wheel, r its radius, and let P be the force which is just able to lift the

FIG. 62.

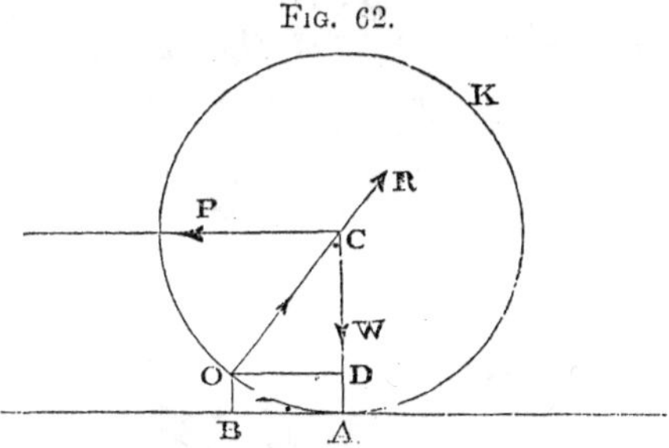

wheel over the obstacle. Let OAK be the wheel just rising from the ground. Let OB be the obstacle over which it is to be drawn, and P the force required. Then, if the wheel be on the point of moving, the three forces in equilibrium are P and W at C, the centre of the wheel, and the reaction R of the obstacle OB acting at O, at right angles to the circumference of the wheel, and therefore likewise passing through C. Now, it will be seen that the sides of the triangle $O\,C\,D$ are respectively parallel to the directions of the forces, and are, therefore, proportional to them. Hence

$$P : R : W :: D\,O : O\,C : C\,D$$

$$\text{or } \frac{P}{W} = \frac{D\,O}{C\,D} = \frac{\sqrt{(r^2 - C\,D^2)}}{C\,D} = \frac{\sqrt{r^2 - (r-h)^2}}{r - h}$$

$$\therefore P = \frac{\sqrt{(2\,h\,r - h^2)}}{r - h} \,.\, W.$$

§ 127. **Polygon of Forces.**—If any number of forces P_1, P_2, P_3, P_4, P_5 . . . . acting at a point O be represented by $A\,B$, $B\,C$, $C\,D$, $D\,E$, $E\,A$, the

Fig. 63.

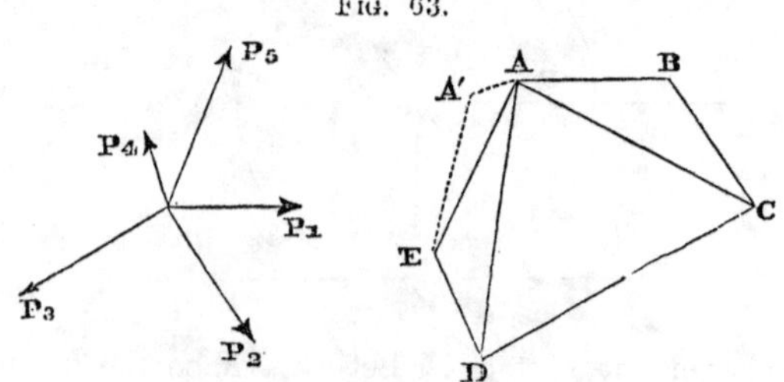

sides of a closed polygon taken in order, the forces shall be in equilibrium; and if the forces be in equilibrium, they shall be capable of being represented by the sides of a polygon.

Join $A\,C$, $A\,D$. Then $A\,C$ represents the resultant of $A\,B$ and $B\,C$; and $A\,D$ represents the resultant of $A\,C$ and $C\,D$, *i.e.* of $A\,B$, $B\,C$, and $C\,D$;

and $A\,E$ represents the resultant of $A\,D$ and $D\,E$, *i.e.* of $A\,B$, $B\,C$, $C\,D$, and $D\,E$; therefore, $E\,A$ together with $A\,B$, $B\,C$, $C\,D$, and $D\,E$ produces equilibrium. Hence P_1, P_2, P_3, P_4, P_5 represented by the sides of the polygon, taken in order, are in equilibrium. The converse may be similarly proved.

If the lines representing the forces do not form a closed polygon, that is, if $E\,A'$ drawn parallel to P_5, and representing it in magnitude, does not meet $A\,B$ in A, the forces are not in equilibrium. In this case the line joining the points $A\,A'$ represents the resultant of the forces, and $A'\,A$ would keep them in equilibrium. In fig. 50, § 116, it will be seen that $O\,A\,a\,b\,c$ is an open polygon, the sides of which represent P_1, P_2, P_3, and P_4, and that $O\,c$ the line closing it represents their resultant.

EXERCISES.

1. Three forces whose magnitudes are as 3 : 2 : 7 act at a point; can they be equilibrium?
2. Three equal forces act at a point; what triangle will represent them?
3. Two forces acting at a point are represented by $A\,B$ the base, and $C\,A$ one of the sides of an isosceles triangle; find the resultant.
4. A string fixed at its extremities to two points in the same horizontal line supports a ring weighing 10 ozs.; the two parts of the string contain an angle of 60°; find the tension.
5. Three forces of 20 lbs., 40 lbs., 50 lbs. act at the same

point, and make angles of 30°, 60°, and 90° respectively with a given straight line; determine their resultant.

6. A weight of 48 lbs. is supported by two strings which are respectively 3 ft. and 4 ft. long, and are fastened to points in the same horizontal line at such distances apart that the strings make right angles with each other; find the tension in each.

7. Show that three forces acting at a point, but not in the same plane, cannot be in equilibrium.

8. $ABCDEF$ is a regular hexagon, and forces represented by the lines AB, AC, AD, AE, AF act at A; find their resultant.

9. Two forces acting at a point are represented by the semi-diagonals AO, OD, of a parallelogram $ADBC$; show which of the sides represents their resultant.

10. Three forces acting at a point are represented by three adjacent sides of a regular hexagon taken in order; find their resultant.

11. A boat is moored to two points on opposite banks, so that the line joining them is perpendicular to the direction of the stream and 15 ft. in length. The two strings make a right angle with each other, and one of them is 12 ft. long, and the force of the stream is equivalent to 20 lbs.; find the tension in each string.

12. Three flexible strings A, B, and C are fastened to a small smooth ring; A and C pass without friction over two fixed pulleys; what weight must be attached to B in order that the forces acting on the ring may be in equilibrium in each of the following cases:—

(*a*) Weight hung to $A = 9$ ozs., weight hung to $C = 9$ ozs.; angle between the strings A and $C = 120°$.

(*b*) Weight hung to $A = 9$ ozs., weight hung to $C = 9$ ozs.; angle between the strings = 180°.

(*c*) Weight hung to $A = 9$ ozs., weight hung to $C = 18$ ozs.; angle between the strings = 150°.

(*d*) Weight hung to $A = 9$ ozs., weight hung to $C = 12$ ozs.; angle between the strings = 90°.

XX. *Resolution of Forces—Analytical Conditions of Equilibrium when any number of Forces act at a Point.*

§ 128. When a single force is replaced by others which together produce the same effect it is said to be resolved, and the several resolved parts are called *components.*

If a weight be pulled along a road by a string inclined to it at an angle, it is evident that a part only of the force employed is used in drawing the weight *along* the road, whilst the remainder tends to lift it *off* the road. That force which produces in a certain direction the same effect as another force acting in a different direction is called its resolved part in that direction.

§ 129. It is evident from the parallelogram of forces that any single force can have two components

Fig. 64.

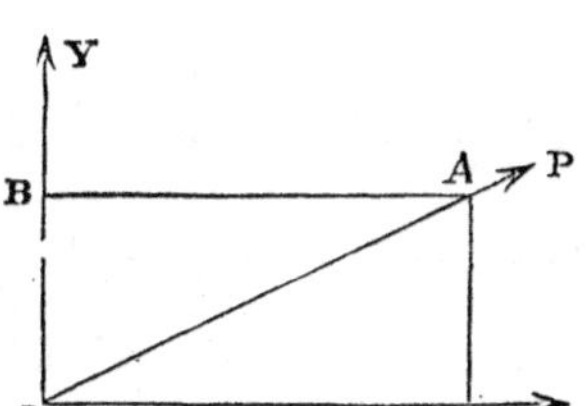

in any directions whatever, since the same straight line may be the diagonal of any number of different parallelograms. The most convenient components into which a force can be resolved are those the directions of which are at right angles to each other. Thus, if P be a force acting at O, and represented by $O A$, then, since $O A$ is the diagonal of the rectangle $O A B C$, the two forces represented by $O B$ and $O C$ produce the same effect as P; and if we call X the force acting along $O C$, and Y the force acting along $O B$, $P^2 = X^2 + Y^2$.

This method of resolution is the simplest, because no part of X aids or opposes Y. Neither component has any part in the other.

§ 130. **The projection of a force in any line represents the resolved part of the force along**

FIG. 65.

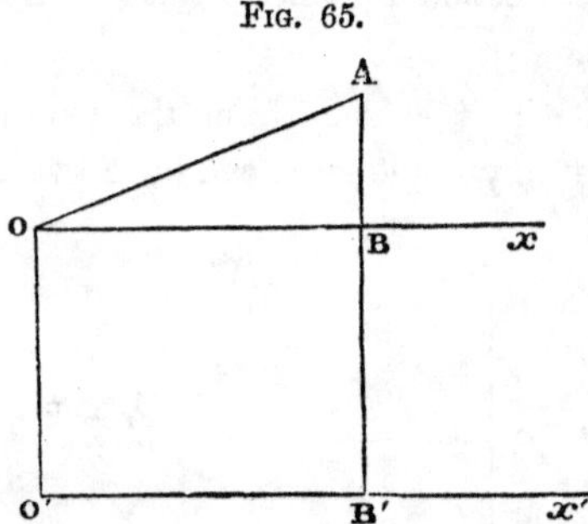

that line.—From the preceding it follows that the resolved part of $O A$ in the direction $O x$ is represented by the line intercepted between O and the foot of the perpendicular let fall from A on the line $O\,x$; and if $O'\,x'$ be drawn parallel to $O\,x$, and $O\,O'$, $A\,B'$ be drawn perpendicular to $O'\,x'$, then $O'\,B' = O\,B$. Hence $O'\,B'$ represents the resolved part of $O\,A$ in the direction $O'\,x'$. But $O'\,B'$ is called the *projection* of $O\,A$ on the line $O'\,x'$. Therefore the resolved part of a force in any line equals the projection of that force on the given line.

§ 131. Forces may very often be compounded by this method of resolution with greater facility than by the parallelogram of forces. For, if each of the several forces acting at a point be resolved along the same two lines, the sum of their components in either direction can at once be found, and the system is thus reduced to two forces at right angles.

§ 132. **To find the resultant of any number of forces acting at a point in different directions and in the same plane.**—Take three forces P_1, P_2, P_3, represented by $O\,A$, $O\,B$ and $O\,C$ respectively. Take $x\,x'$, $y\,y'$, two straight lines at right angles through O. Project $O\,A$, $O\,B$, and $O\,C$ on each of these lines. Then the three forces may be replaced by $O\,L$ and $O\,L'$, by $O\,M$ and $O\,M'$, by $O\,N$ and $O\,N'$. But $O\,L$, $O\,M$, and $O\,N$ act in the same straight line, and therefore their resultant X is their algebraical sum. Also, $O\,L'$, $O\,M'$, $O\,N'$ act in one straight line and have a resultant y equal to their algebraical sum. Thus the original forces P_1, P_2, and P_3 are equivalent

FIG. 66.

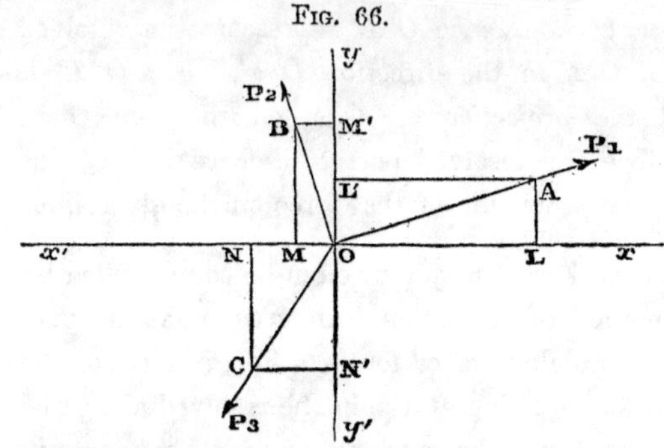

to X and Y at right angles, where

$$X = OL - OM - ON;$$

and $$Y = OL' + OM' - ON';$$

and if R be the resultant of X and Y, *i.e.* of all the forces,

$$R^2 = X^2 + Y^2.$$

In the same way the resultant of any number of forces may be obtained. Hence, if any number of forces act at a point, and two lines be drawn through this point at right angles to each other, and if X and Y be the algebraic sums of the components of these forces along these lines respectively, then $R^2 = X^2 + Y^2$, where R is the resultant of all the forces.

This method of finding the resultant of any number of forces exemplifies the scientific method known as *analysis and synthesis*. The method we before considered, which consisted of finding the resultant of two forces and combining this with a third force, and the new resultant with a new force continually, was an illustration of synthesis only.

§ 133. **Examples.**—(1) Three forces of 4, 5, and 6 lbs. act at a point. The angle between the

FIG. 67.

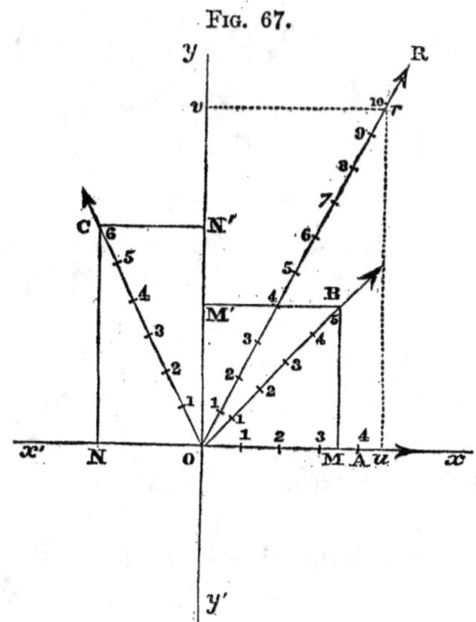

first and second is 45°, and between the second and third 75°. Find their resultant.

It is generally convenient to take the line of action of one of the forces as the line Ox. Draw Oy at right angles to it, through O. Project the forces along Ox and Oy. Then if OA, OB, OC represent P_1, P_2, and P_3 we have

$$X = OA + OM - ON$$

$$= 4 + \frac{5}{\sqrt{2}} - 3 = 1 + \frac{5}{\sqrt{2}};$$

$$Y = OM' + ON' = \frac{5}{\sqrt{2}} + 3\sqrt{3}$$

and
$$R^2 = \left(1 + \frac{5}{\sqrt{2}}\right)^2 + \left(\frac{5}{\sqrt{2}} + 3\sqrt{3}\right)^2$$
$$= \tfrac{1}{2}\{(\sqrt{2} + 5)^2 + (5 + 3\sqrt{6})^2\}$$
$$= 53 + 5\sqrt{2} + 15\sqrt{6}$$

$$\therefore R = 10 \text{ nearly.}$$

If Ou be taken equal to $OA + OM - ON$, and Ov be taken equal to $OM' + ON'$; and if ur, vr be drawn parallel to Oy and Ox respectively Or will represent the resultant, and this line will be found to measure a little less than ten units.

(2) To find the resultant of equal forces when the angle between them is 15°.—Let the line Ox be so taken that it makes an angle of 30° with one force and 45° with the other, in which case the angle between the forces will be the difference of 45° and 30°, that is, 15°. Draw Oy at right angles to Ox.

The projections of P_1 are $\dfrac{P_1}{\sqrt{2}}$ and $\dfrac{P_1}{\sqrt{2}}$

" " P_2 are $\dfrac{P_2\sqrt{3}}{2}$ and $\dfrac{P_2}{2}$.

But $P_1 = P_2$

$$\therefore X = o\,u = P\left(\frac{1}{\sqrt{2}} + \frac{\sqrt{3}}{2}\right)$$

and $$Y = o\,v = P\left(\frac{1}{\sqrt{2}} + \frac{1}{2}\right)$$

$$\therefore R^2 = X^2 + Y^2 = \frac{P^2}{2}(4 + \sqrt{2} + \sqrt{6})$$

$$\therefore R = P(1{\cdot}98 \cdots)$$

FIG. 68.

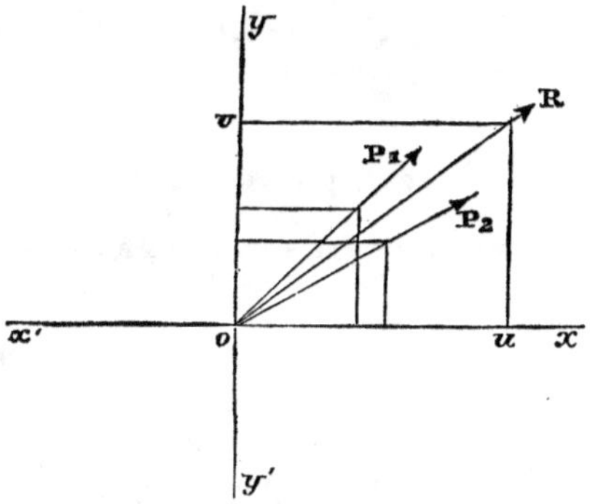

§ 134. **Conditions of Equilibrium when any number of Forces act at a Point.**—If the forces are in equilibrium it is evident that the resultant must be zero; and since $R^2 = X^2 + Y^2$, it follows that $X^2 + Y^2 = 0$ when the forces are in equilibrium, and consequently $X = 0$ and $Y = 0$. Hence, *if any number of forces acting at a point are in equilibrium the algebraical sums of the components of these forces, along any two straight lines drawn at right*

angles to each other through this point, must separately vanish.

The conditions of equilibrium thus found are called the analytical conditions, because the forces are analysed into their components. Remembering what X and Y represent, these two conditions will be indicated by the equations

$$(1)\ X = 0$$
$$(2)\ Y = 0.$$

EXERCISES.

1. Two equal forces acting at a point make angles of 30° and 45° with opposite sides of the same straight line; find their resultant.
2. Two forces of 6 lbs. and 10 lbs. acting at a point include an angle of 105°; find the resultant.
3. Two forces of 8 lbs. and 10 lbs. act at right angles; find the resultant by resolving each along two directions at right angles, one of which makes an angle of 30° with one of the forces.
4. Three forces of 4, 5, and 6 ozs. act at a point and include angles of 60° and 75°; find their resultant.
5. Three equal forces act at a point, and the angle included between the first and second is 30°, between the second and third 105°; find the resultant.
6. Five forces of 5, 2, 4, 3, and 6 ozs. act at a point and include angles of 45°, 75°, 60°, and 90°; find the magnitude of the force which will keep them in equilibrium.
7. Show why the traces of a horse ought always to be parallel to the road along which he is pulling.

8. A man pulls a weight by means of a rope along a road with a force of 20 lbs. The rope makes an angle of 30° with the road. Find the force he would need to apply parallel to the road. Does the inclination of the road alter the answer?

9. Find the magnitude of the force which will keep in equilibrium two forces of 2 lbs. and 4 lbs., (1) at an angle of 15°, (2) at an angle of 75°.

10. The resultant of two forces acting at a point is 4 lbs., and the angle between them is 150°. One of the components is 4 lbs.; find the other.

11. Let ABC be a triangle, and lines be drawn from A, B, C to the middle points of the opposite sides. If three forces acting from A, B, C, respectively be represented by these three lines the forces will be in *equilibrium*.

12. Three forces act at a point, and include angles of 90° and 45°. The first two forces are each equal to $2P$, and the resultant of them all is $\sqrt{10}\,P$; find the third force.

13. Indicate the forces that maintain a kite in equilibrium.

14. Find the resultant of three forces, the least of which is 10 lbs., which are represented by, and act along OA, OB, OC, two sides and the diagonal of an oblong whose area is 60 square inches, and shorter side 5 ins.

15. A man and a boy pull a heavy weight by ropes inclined to the horizon at angles of 60° and 30° with forces of 80 lbs. and 100 lbs. The angle between the two vertical planes of the cords is 30°; find the single horizontal force that would produce the same effect.

16. The resultant of two equal forces is 56 lbs., and the included angle is 15°; find the forces.

17. A weight of 10 lbs. is supported by two strings, one of which makes an angle of 30° with the vertical, the other 45°; find the tension in each string.

18. Three forces each of 10 lbs. act at the same point; the first makes an angle of 30° with the second, and the

second an angle of 60° with the third; calculate the magnitude of the single force which with them will produce equilibrium.

XXI. *Forces not meeting at a Point—Parallel Forces.*

§ 135. Parallel forces are those acting at different points of a body and in directions parallel to one another. If they act in the same direction they are called *like* forces; if in opposite directions, *unlike*. A pair of horses harnessed to a carriage is an example of two like parallel forces.

§ 136. **Resultant of two Parallel Forces.**—If two forces are parallel and act in the same direction, it is evident that their joint effect is the sum of their separate effects; and if they act in opposite directions it is equal to their difference.

The magnitude of the resultant of parallel forces is, therefore, the same as of forces in the same straight line, *i.e.*, is equal to their algebraical sum.

It has been shown (§ 121) that when two forces act at a point their resultant is nearer to the greater force. If the forces act in like directions, from or towards the same parts, the resultant will be found to act *between* them; if in unlike directions, it will be found to act *outside* both forces, as shown in the annexed diagram (fig. 69). This proposition is equally true when the forces are parallel, *i.e.*, when the point O is infinitely distant. But when the forces are parallel the nearness of the resultant to either of the forces can be more easily estimated than when the forces are inclined at an angle, since it can be exactly measured by the perpendicular distance of the direction of the resultant from that of each of the

Fig. 69.

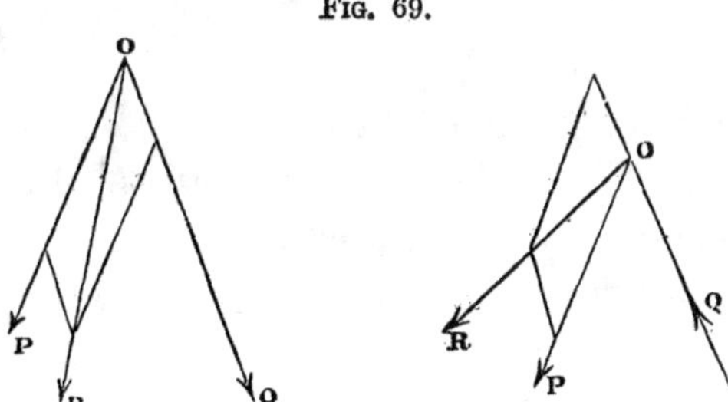

forces. In parallel forces it is found that the resultant is as much nearer to the greater force as that force is greater than the lesser force, and acts between like and outside unlike parallel forces. Hence parallel forces are only a particular case of those which we have already considered.

If P and Q be two parallel forces, *like* in fig. 70, i., and *unlike* in fig. 70, ii., the resultant R will act between P and Q in fig. i., and outside them in fig. ii.; but in both cases, if P is greater than Q, R will be as much nearer to P as P is greater than Q; *i.e.*, R's perpendicular distance from P will be as many times less than R's perpendicular distance from Q as P is greater than Q. Hence, if $m\,C\,n$ be perpendicular to their directions,

i. Fig. 70. ii.

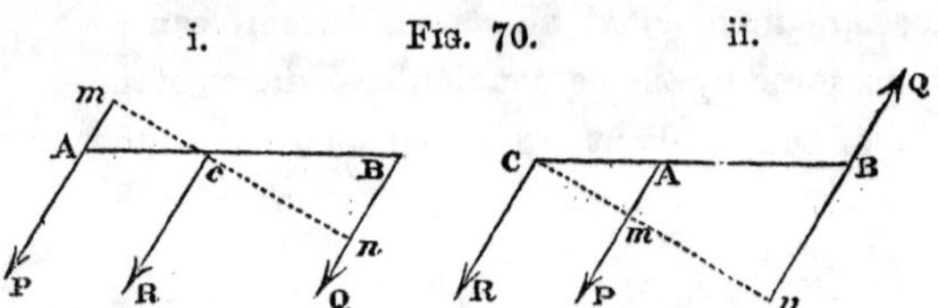

$$P : Q :: Cn : Cm$$

and $\therefore P : Q :: CB : CA$ (by similarity of triangles).

Hence the magnitude and position of the resultant is determined by the equations

$$R = P \pm Q$$

and $$P \times CA = Q \times CB.$$

§ 137. **Proof.**—The position of the resultant is thus explained. We now proceed to prove the above proposition.

I. *Like Forces.*—Suppose the two forces to be acting at A and B, and to be represented by $A\,a$ and $B\,b$. At A apply any force X in the direction $B\,A$, and at B an equal force X' in the direction $A\,B$. These forces will produce no effect, since they are equal and opposite.

Draw the diagonals $A\,s$, $B\,s'$. Then $A\,s$ represents S, the resultant of P and X; and $B\,s'$ represents S', the resultant of Q and X'. These two pairs of forces can therefore be replaced by S and S'. If the lines $A\,s$, $B\,s'$ be produced they will meet at some point O, since the angles $X A\,s$, $X' B\,s'$ are together less than two right angles; and the forces S and S' may therefore be transferred to O.

Draw $O\,C$ parallel to $A\,a$ or $B\,b$, and resolve S

FIG. 71.

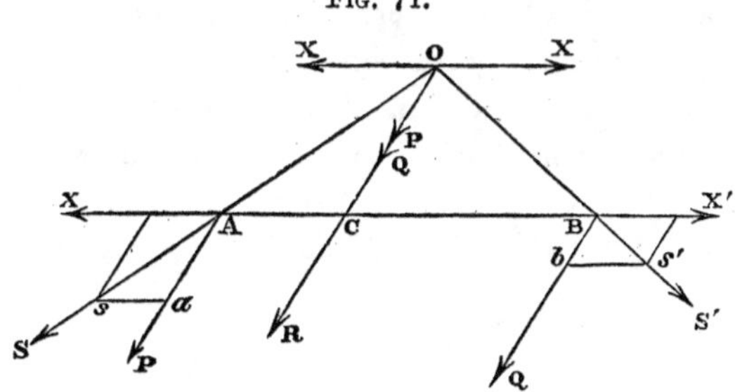

into its two components, one along $O\,C$ and the other parallel to $A\,B$, and let S' be similarly resolved. Then we have X and X' acting at O in opposite directions, and counteracting one another; and P and Q acting along $O\,C$. Hence $P + Q$ at C produces the same effect as P at A and Q at B, and therefore the resultant of P and Q acts at C parallel to their direction. Let R be the resultant; then $R = P + Q$.

Now, the triangle $O\,C\,A$ has its sides parallel to the directions of P, X, and S,

and $$\therefore \frac{P}{X} = \frac{O\,C}{C\,A}$$

Similarly $$\frac{X'}{Q} = \frac{C\,B}{O\,C}$$

$$\therefore \quad \frac{P}{Q} = \frac{C\,B}{C\,A} \text{ since } X = X'$$

or $$P \times C\,A = Q \times C\,B.$$

II. *Unlike Forces.*—Let P and Q be two unlike forces acting at A, and B, and let Q be greater than P. At A, and B apply two equal forces X and X' in the directions $B\,A$ and $A\,B$. Then $A\,s$ will represent S, the resultant of P and X, and $B\,s'$ will represent S', the resultant of Q and X'. Let $A\,s$ and $B\,s'$ meet at O. Then S and S' act at O, and

Fig. 72.

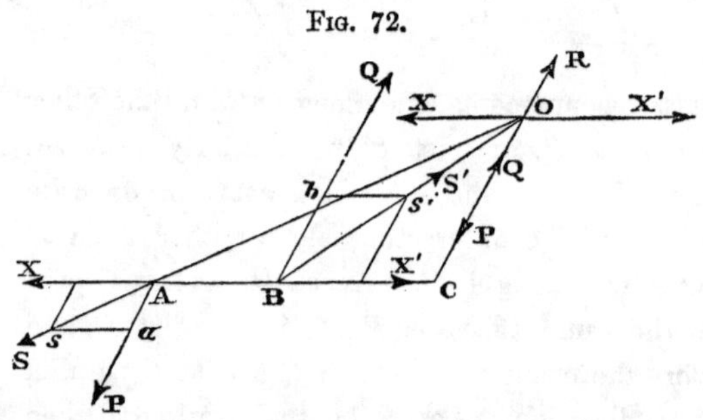

may be resolved into their component parts, viz., X and X' parallel to $A\,B$ destroying one another, and P acting along $O\,C$ and Q in the direction of $C\,O$. Hence P at A and Q at B are equivalent to a force R equal to $Q - P$ in the direction $C\,O$.

Therefore the resultant of P and Q is equal to $Q-P$, and acts parallel to the original forces at the point C.

Also, since the sides of the triangle $A\ C\ O$ are parallel to the directions of P, X, and S, and the sides of the triangle $B\ C\ O$ are parallel to the directions of Q, X', and S', we have

$$\frac{P}{X}=\frac{O\,C}{C\,A} \text{ and } \frac{X'}{Q}=\frac{B\,C}{C\,O}$$

and $\therefore$ $$\frac{P}{Q}=\frac{B\,C}{C\,A} \text{ since } X=X'$$

or $$P\times C\,A=Q\times B\,C.$$

§ 138. **Experimental Verification.**—The above propositions may be experimentally verified in the following way:—

Let $X\,X'$ be a rod suspended at C. Let the two

FIG. 73.

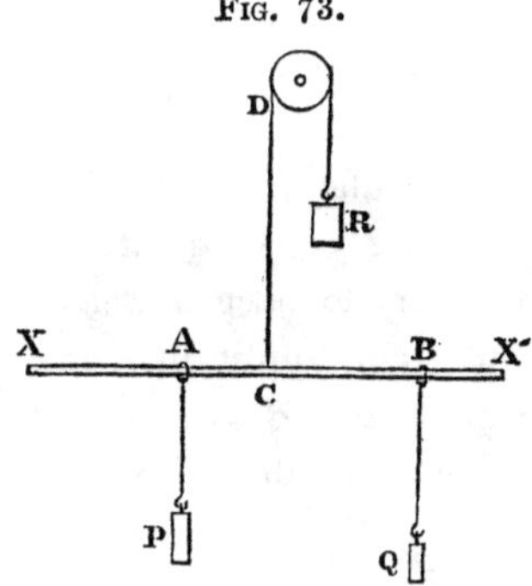

weights P and Q hang from any two points A and B (fig. 73), and let a weight R equal to their sum be attached to the end of a thin cord that passes without friction over the pulley D. Then if the weights P and Q be moved till they are in equilibrium, and if the distance CA be equal to p, and the distance CB equal q, it will be found that $P \times p = Q \times q$.

Again, let the weight P hang from A, and let the weight Q be fastened to the end of the string

FIG. 74.

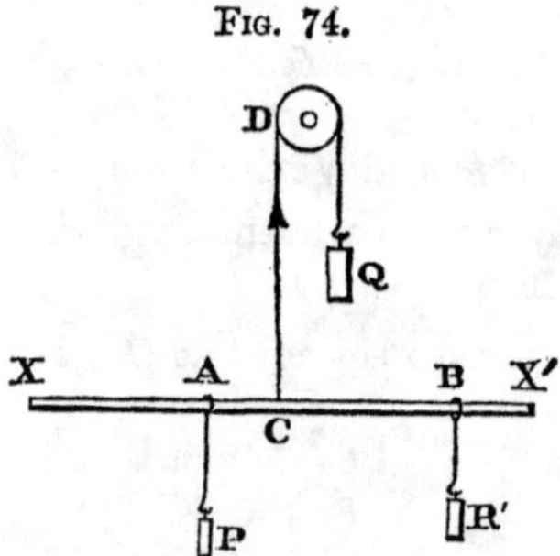

that passes over the pulley. Then P and Q are two *unlike* forces, and if Q be greater than P, it will be found necessary to hang a weight R' from some point B to maintain equilibrium, and if $R' = Q - P$, and the distance $AB = p$ and the distance $CB = q$, it will be found that $P \times p = Q \times q$ as before.

§ 139. **Examples.**—(1) If $P = 4$ lbs., and $Q = 5$ lbs. be two like parallel forces, and the distance between their points of application be 12 ins., find the position of the resultant.

Let x equal R's distance from P,
then $12-x$ equals „ „ „ Q
$\therefore 4 \times x = 5\,(12 - x) = 60 - 5\,x$
$\therefore x = 6\frac{2}{3}$ and $12 - x = 5\frac{1}{3}$.

(2) If the resultant of two unlike parallel forces is 12 lbs., and acts at a distance of 5 ins. from the greater and 7 ins. from the lesser force, find the forces.

Let P be greater than Q, then $R = P - Q = 12$
and $\therefore P = 12 + Q$
also $5 \times P = 7 \times Q$
$\therefore 5 \times (12 + Q) = 7 \times Q$
$\therefore Q = 30$ lbs. and $P = 42$ lbs.

(3) The resultant of two like forces is 10 lbs., and is twice as near to P as to Q; find the forces.

$$P + Q = 10 \therefore Q = 10 - P.$$

Let x be P's distance from R.
Then $P \,.\, x = (10 - P) \,.\, 2\,x$
$\therefore P = 20 - 2\,P$
$\therefore P = 6\frac{2}{3}$ lbs. and $Q = 3\frac{1}{3}$ lbs.

§ 140. **To find the Position of the Resultant of a number of Parallel Forces acting at different points of a rigid body.**—Let P_1, P_2, P_3 . . . be parallel forces acting at A, B, C . . . Then, the resultant R_1, of P_1 and $P_2 = P_1 + P_2$, and acts at a point O_1, such that $P_1 \times A\,O_1 = P_2 \times B\,O_1$. Similarly R_2, the resultant of R_1 and P_3, acts at a point O_2, such that $R_1 \times O_1\,O_2 = P_3 \times C\,O_2$.

or, $$(P_1 + P_2) \times O_1\,O_2 = P_3 \times C\,O_2$$

and $$R_2 = P_1 + P_2 + P_3.$$

Fig. 75.

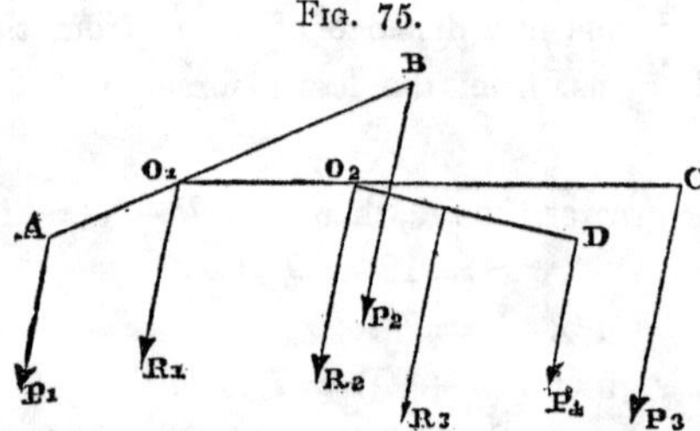

In the same way, by combining this resultant with a new force, and the resultant of these with another force, the resultant of any number of parallel forces may be obtained.

§ 141. **Centre of Parallel Forces.**—The point at which the resultant of any number of parallel forces acts is called the *centre* of the forces. It is evident from the foregoing investigation that the position of the centre of any number of parallel forces is independent of the directions of the forces, and depends only on their points of application and their respective magnitudes.

The position of this point may be easily found in most cases by the following method :—

If P_1 and P_2 be the forces, and O any point; and if through O any line be drawn cutting their directions in the points A and B and their resultant at C, then $P_1 \times OA + P_2 \times OB = R \times OC$.

Since $P_1 \times AC = P_2 \times BC$, it follows that

$$P_1 \times (OA - OC) = P_2 \times (OC - OB)$$

$$\therefore P_1 \times OA + P_2 \times OB = (P_1 + P_2) \times OC = R \times OC.$$

Fig. 76.

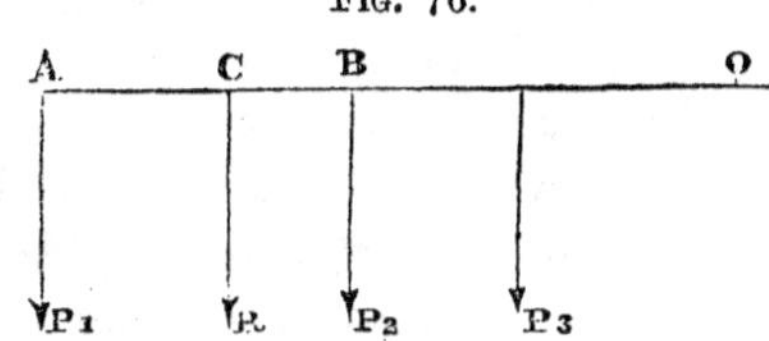

Similarly by combining R with P_3, and that resultant with a new force, we can prove that if P_1, P_2, P_3 . . . be any number of parallel forces, x_1, x_2, x_3 . . . their respective distances from a point O, along any straight line drawn through that point, and x the distance of their resultant from the same point, then $Rx = P_1 x_1 + P_2 x_2 + P_3 x_3 + \ldots$

and $$\therefore x = \frac{P_1 x_1 + P_2 x + P_3 x_3 + \ldots}{P_1 + P_2 + P_3 + \ldots}.$$

§ 142. **Equilibrium of Parallel Forces.**—If P_1, P_2, P_3 . . . be forces in equilibrium, some acting in one direction, and some in the opposite direction, their resultant equals zero, and the condition of equilibrium is therefore given by the equation

$$P_1 x_1 + P_2 x_2 + P_3 x_3 + \ldots = 0.$$

In this equation forces pulling in opposite directions are supposed to have opposite signs; and if the point O, through which the line of reference passes, and which may be taken anywhere, lies between the forces, some being on one side of it and some on the other, a further distinction of sign is necessary, and all the *distances* on one side of O must be accounted positive, and those on the other side negative. Bearing in mind these distinctions of sign, the condition of equilibrium may be thus stated: *Parallel forces are in equilibrium, when the algebraical sum of the products of the forces into their respective distances, from some fixed point, measured along a line drawn through that point across their directions equals zero.*

This proposition is most important in the solution of problems. It is often found convenient to take the point O in the line of one of the forces, in which case the distance of that particular force from O vanishes.

§ 143. **Examples.**—(1) A beam AB, 10 ft. long, the weight of which is 10 lbs., and acts at its middle point, is supported on two props, 1 ft. and 2 ft. from the ends A and B respectively. From the extremity A a weight of 20 lbs. hangs; find the pressure on each prop.

Here we have two forces acting vertically downwards, viz. 10 lbs and 20 lbs., and the reactions of

FIG. 77.

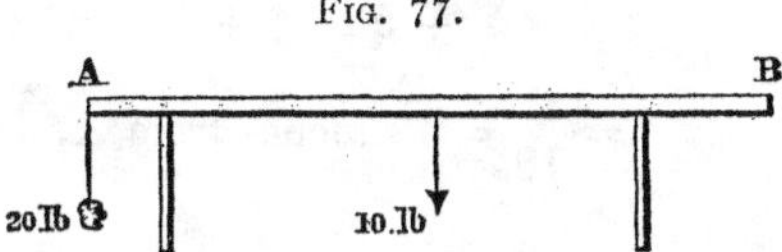

the two props, which are together equivalent to these forces, acting vertically upwards.

Let P and Q be the pressures on the two props.

Then $P + Q = 30$. Let the fixed point from which the distances are to be measured be at A. Then we have

$$P \times 1 + Q \times 8 - 10 \times 5 - 20 \times 0 = 0$$
$$\therefore (30 - Q) + 8\,Q = 50$$
$$7\,Q = 20 \;\therefore\; Q = 2\tfrac{6}{7} \text{ and } P = 27\tfrac{1}{7}.$$

(2) Weights of 3 ozs., 4 ozs., and 5 ozs. are hung on a rod $A\,B$, 12 inches long, at distances of 1 inch, 2 inches, and 8 inches from the end A (fig. 78). If the weight of the rod be 6 ozs. and act at its centre, find where the rod must be suspended, that it may rest horizontal.

Here we have a force of 3 oz. at a distance of 1 inch from A,

a force of 4 ozs. at a distance of 2 inches from A

„ „ 5 „ „ „ 8 „ „

and „ „ 6 „ „ „ 6 „ „

The only force acting in the opposite direction is the tension of the string, which is equal to $3 + 4 + 5 + 6 = 18$ ozs., and is fixed to the rod at a distance x to be found.

$$\therefore\ 18x = 3 + 4 \times 2 + 5 \times 8 + 6 \times 6 = 87$$

$$\therefore\ x = \frac{87}{18} = 4\tfrac{5}{6} \text{ inches from } A.$$

Fig. 78.

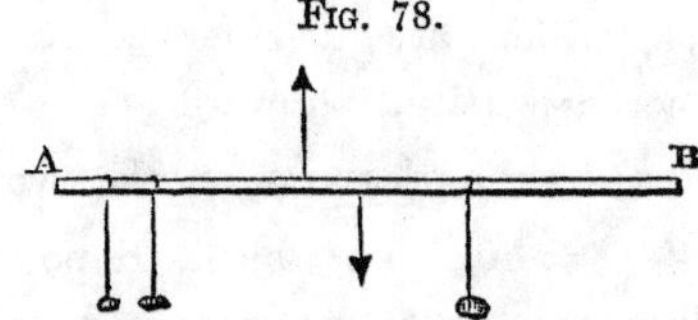

(3) Four like parallel forces, the magnitudes of which are 2 ozs., 3 ozs., 4 ozs. and 5 ozs., act at the four corners of a square; find the position of their resultant.

It is sometimes convenient to combine, first one pair of forces, and then another pair, and finally to combine the two resultants.

The resultant of the forces 3 and 4 is a force of 7 ozs. acting at O_1 (fig. 79), where $B\,O_1$, equals $\frac{4}{7}\,B\,C$. The resultant of 2 and 5 is a force of 7 ozs. acting at O_2, where $A\,O_2$ is $\frac{5}{7}$ of $A\,D$. Join $O_1\,O_2$ and the final resultant will be a force of 14 ozs. acting at the point M which bisects $O_1\,O_2$. Hence, the centre of the forces is situated at a point equally distant from

$A\,D$ and $B\,C$ and $\frac{9}{14}$ of one of the sides from $A\,B$ and $\frac{5}{14}$ from $D\,C$.

Fig. 79.

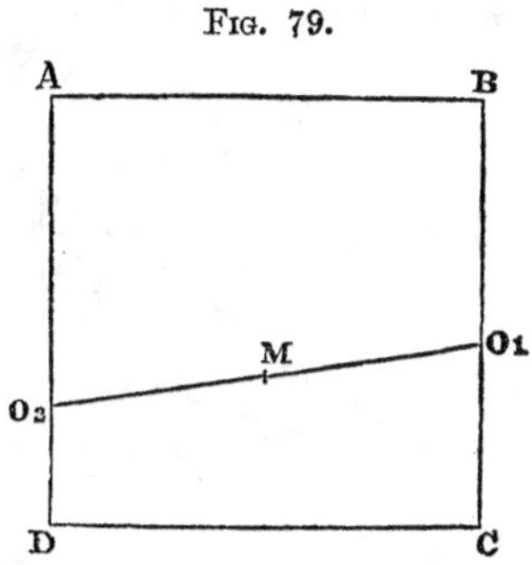

§ 144. **Resolution of a force into Parallel Components.**—A force F may be resolved into two parallel components P and Q, provided that $P : Q :: Q$'s distance from $F : P$'s distance from F; and either or both of these components may be again resolved, provided that the relation between the distances of the components from the original force is preserved.

§ 145. **Example.**—A weight W rests on a triangular table at a point in the line joining one of the angular points with the middle of the opposite side, and at a distance from the angular point equal to twice its distance from the side. The table is supported on three props at its angular points; find the pressure on each prop.

Let W at E be resolved into two parallel forces at D and C. Then, since $CE = 2\,ED$, the component at D equals twice the component at C, and

FIG. 80.

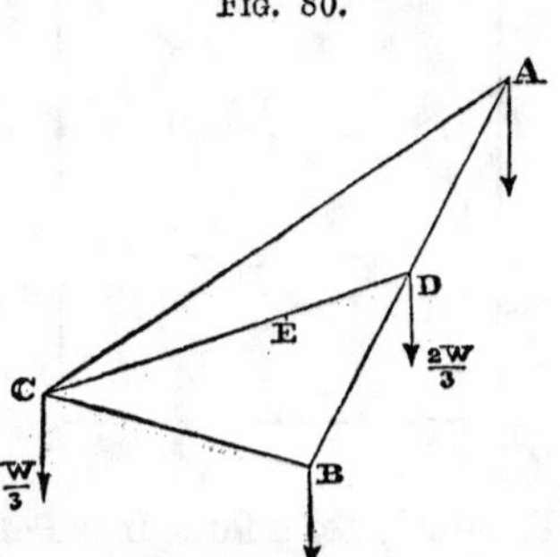

$\frac{2\,W}{3}$ acts at D and $\frac{W}{3}$ acts at C. But $\frac{2\,W}{3}$ at D is equivalent to $\frac{W}{3}$ at A and $\frac{W}{3}$ at B; therefore, the weight W exerts a pressure of $\frac{W}{3}$ on each prop.

§ 146. **Couples.**—In § 137, II. we found the resultant of two *unlike* parallel forces P and Q, on the supposition that Q was *greater* than P. Let us see what would have happened if Q had been *equal* to P. In this case $A\,s$ and $B\,s'$ (fig. 72) would have been parallel, and the point O through which the resultant passes would have been infinitely distant. It thus appears that two equal and opposite parallel forces have no resultant, *i.e.*, there is no single force tending to produce translation that can replace them. Such a pair of forces is called a *couple*, and the perpendicular distance between the forces is called their arm. The effect of a couple is to produce *rotation*, and it generally happens that one of the forces is replaced by the reaction of a fixed point about which the body is free to rotate. The conditions of equilibrium of a body free to rotate about a fixed point will be considered in the next Lesson.

EXERCISES.

1. Parallel forces are applied at two points 5 ins. apart, and are kept in equilibrium by a third force 3 ins. from the one, and 2 ins. from the other. What is the ratio of the forces?
2. Two men, of the same height, carry on their shoulders a pole 6 ft. long, and a weight of 121 lbs. is slung on it, 30 ins. from one of the men; what portion of the weight does each man support?
3. A beam the weight of which is equivalent to a force of 10 lbs. acting at its middle point is supported on two props at the end of the beam. If the length of the beam be 5 ft., find where a weight of 30 lbs. must be placed, so that the pressure on the two props may be 15 lbs. and 25 lbs. respectively.
4. Two weights of 3 ozs. and 5 ozs. hang at the ends of a rod 12 ins. long, and a third weight of 6 ozs. is placed 3 ins. from the lighter weight; find the position of the resultant.
5. Equal weights hang from the corners of a triangle; find the point at which it must be suspended to rest horizontally.

6. Equal forces act along AB, BC, DC, AD, the sides of a square; find the resultant.
7. The ratio of two unlike parallel forces is $\frac{4}{5}$, and the distance between them is 10 ins.; find the position of the resultant.
8. The resultant of two unlike forces is 6 lbs., and acts 8 ins. from the greater force, which is 10 lbs.; find the distance between them.
9. Forces of 3, 4, 5, 6 lbs. act at distances of 3 ins., 4 ins., 5 ins., 6 ins. from the end of a rod; at what distance from the same end does the resultant act?
10. If a weight rest in the middle of a square rough table, will the pressure on each leg be altered if one pair of legs be longer than the opposite pair?
11. A circular table rests on one central leg, and a weight of 10 lbs. is placed midway between the centre and circumference; where must a weight of 8 lbs. be placed so as to preserve equilibrium?
12. A plank weighing 10 lbs. rests on a single prop at its middle point; if it be replaced by two others on each side of it, 3 ft. and 5 ft. from the middle point, find the pressure on each.
13. To a weightless rigid rod 30 ins. long, weights of 4, 6, 8, 12 lbs. are fixed at equal distances from one another; where must the rod be suspended to rest horizontally?
14. A weightless rod is suspended at a point 3 ins. from one end, and a weight of 10 lbs. is hung from the same end; if the rod be 15 ins. long, find the weight that must be put at the other end to maintain equilibrium.
15. If the rod be heavy, and its weight be equivalent to a vertical force of 2 lbs. acting at its middle point, what weight must then be placed at the further end?
16. Four vertical forces of 4, 6, 7, 9 lbs. act at the four corners of a square; find their resultant.
17. Find the *centre* of like parallel forces, 3, 2, 5, 7 lbs., which act at equal distances apart along a straight rod 12 ins. long.

18. A horizontal straight bar 6 ft. long and weighing 16 lbs. is supported at each end, and a weight of 48 lbs. is hung at 2 ft. from one end; find the pressure upon each of the supports.
19. Three parallel forces of 4 lbs., 5 lbs., and 7 lbs. are applied respectively at the centre and two extremities of a rigid bar; find their resultant in magnitude and position.
20. A flat board 12 ins. square is suspended in a horizontal position by strings attached to its four corners *A B C D*, and a weight equal to the weight of the board is laid upon it at a point 3 ins. distant from the side *A B* and 4 ins. from *A D*; find the relative tensions in the four strings.

XXII. *Forces producing Rotation—Moments.*

§ 147. We have hitherto considered forces, the tendency of which has been to produce translation, or motion from one place to another; we have now to treat of the tendency of forces to produce rotation about some fixed point.

If a point in a body be supposed fixed, so that the body cannot move out of its own place, but is free to rotate about that point, a force applied at any other point and in a direction that does not pass through the fixed point will produce *rotation*.

§ 148. The rotatory effect of a force depends on:—

First. The magnitude of the force.

Secondly. The perpendicular distance of its direction from the *fixed* point.

In closing a door a small force applied at the handle will produce the same effect as a much larger force applied at a point nearer to the hinge.

§ 149. **Moment of a Force.**—The tendency of a force to produce rotation about a fixed point is called its *moment* about that point, and is measured by the

FIG. 81.

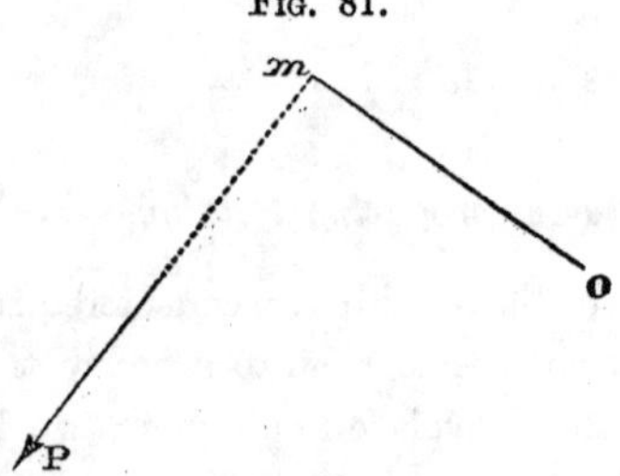

product of the number of units of force into the number of units of length in the perpendicular drawn from the fixed point on to the direction of the force. Thus if P be a force tending to cause a body to rotate about the point O, and $O\,m$ be the perpendicular drawn from O on to its direction, then $P \times O\,m$ measures the rotatory tendency of the force, and is called the moment of the force P about the point O.

§ 150. **Geometrical Measure of the Moment of a Force.**—If $A\,B$ represent the magnitude of P, then the measure of the moment is $AB \times Om$, which is equal to twice the area of the triangle AOB. Hence the moment of a force about a point may be measured by twice the area of the triangle which

FIG. 82.

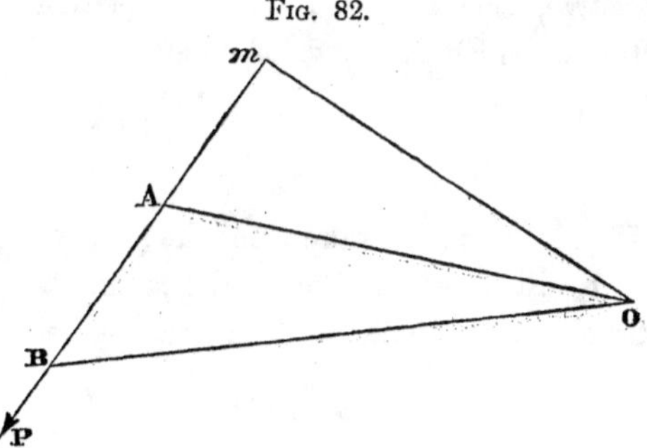

has the straight line representing the force for a base, and the fixed point for an apex.

If two triangles having the same apex can be shown to be equal, the moments of the forces, represented by the bases, about that apex will be equal also.

§ 151. The moment of a force about a point in its own line of action is evidently nothing, since no perpendicular can be let fall on a line from a point in the same line.

§ 152. **Positive and Negative Moments.**—In considering the action of forces we found it convenient to distinguish between positive and negative forces, according as they tended to move the body to the right or to the left of a fixed point. In the same way it is found desirable to speak of positive and negative moments. The moment of a force is said to be positive, when it tends to produce rotation in the direction in which the hands of a clock move, and negative when its tendency is in the reverse direction.

§ 153. **If any point be taken in the line of action of the resultant of two Forces the Moments of the Forces about this point will be equal.**—Let

Fig. 83.

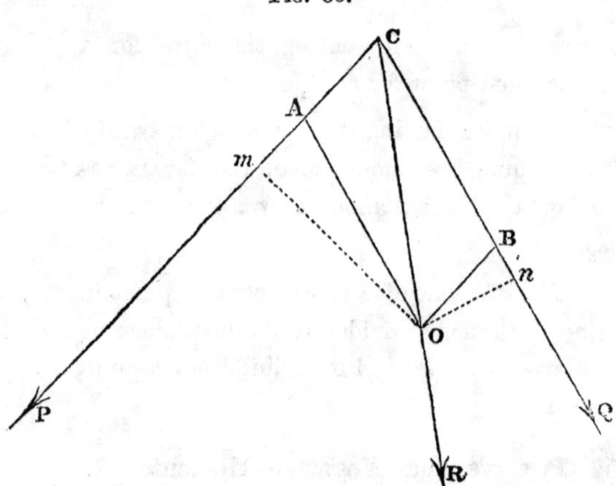

P and *Q* be two forces acting at *C*, and let *R* be their resultant. In the line of action of *R* take any point O, and draw OA, OB parallel to the directions of Q and P respectively. Then CO, CA, and CB will represent R, P, and Q, wherever the point O may be taken. But the moment of the force CA about the point O is represented by twice the area of the triangle CAO (§ 150), and the moment of CB by twice the triangle CBO, and

$$\text{the triangle } CAO = \text{the triangle } CBO.$$

$\therefore$ The moment of P about O equals the moment of Q about the same point.

$$\text{Or,} \qquad P \times Om = Q \times On.$$

§ 154. **Equilibrium of two Moments.**—If two forces act at different points of a body which is free to rotate about a fixed point they will produce equilibrium when their moments about that point are equal and opposite. Now, we have seen that the moments of two forces are equal and opposite about any point in the line of action of their resultant. Hence a body acted upon by two forces will be in equilibrium when the point of support is in the line of action of the resultant.

§ 155. **Application to Lever.**—The lever has been defined as a rigid rod capable of turning about a fixed point called the Fulcrum. We are now able to find the conditions of equilibrium on the lever, and the relation between the forces acting upon it in other cases than those already considered. For if P and Q be two forces acting at the points A and B of a rigid rod of any form, and if the rotatory tendencies of these forces are equal about the point C or fulcrum, then the resultant of P and Q passes through C, and the moment of P about C is equal and opposite to the moment of Q about the same

Fig. 84.

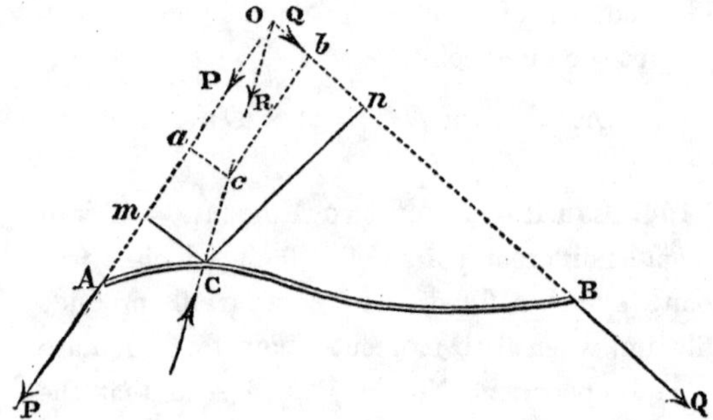

point. Hence the following propositions hold good with respect to the lever, whether the lever be straight or bent, and whether the forces are perpendicular to the arms or not.

(1) The resultant of the two forces passes through the fulcrum of the lever, and is equal to the pressure on the fulcrum.

(2) The algebraical sum of the moments of the forces about the fulcrum equals zero.

This latter proposition is generally known as the *principle* of the lever.

In fig. 84 we see that if the directions of P and Q be produced to O, and $O\,a$ and $O\,b$ represent the two forces, and $O\,c$ their resultant, and if $O\,c$ be produced to cut $A\,B$ in C, then C is the fulcrum, and the pressure of the fulcrum is represented by $O\,c$, and

$$P \times C\,m = Q \times C\,n$$

where $C\,m$ and $C\,n$ are perpendiculars drawn from C to the directions of P and Q.

§ 156. **Examples.**—(1) A bent lever without weight consists of two arms, one of which is twice as

Fig. 85.

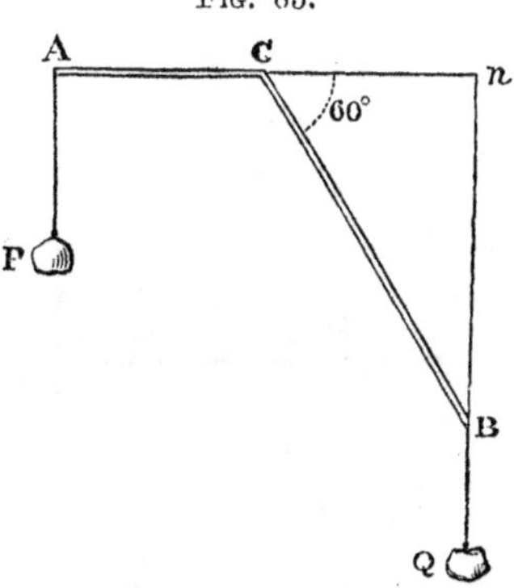

long as the other, and inclined to each other at an angle of 120°; find the ratio of the weights that must be suspended from their ends, so that the lever may rest with the shorter arm horizontal.

Let P and Q be the two weights. Taking moments about C, we have

$$P \times A\,C = Q \times C\,n;$$

and since the angle $B\,C\,n$ is $60°$

$$C\,n = \tfrac{1}{2}\,C\,B = A\,C$$

$$\therefore P \times A\,C = Q \times A\,C \text{ or } P = Q.$$

(2) A rod the weight of which is 10 lbs. and acts at its middle point moves at one end about a hinge

FIG. 86.

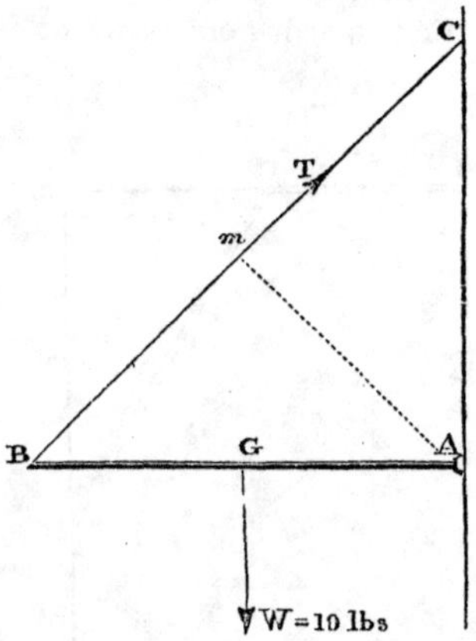

and is supported at the other end by a piece of string attached to a point, vertically over the hinge, and at a distance from it equal to the length of the rod; find the *tension* in the string when the rod rests in a horizontal position. Let $A B$ be the rod moveable about a hinge at A. Let $C B$ be the string, then $A C = A B$. Let W be the weight of the rod at G, and let T be the tension in $B C$.

Then in equilibrium the moment of W about A must be equal to the moment of T about A.

Or, if $A m$ be drawn perpendicular to $C B$

$$T \times A m = W \times A G$$

and $$A m = \frac{A B}{\sqrt{2}}, \text{ and } A G = \frac{A B}{2}$$

$$\therefore \frac{T}{\sqrt{2}} = \frac{W}{2} \text{ or } T = 5 \sqrt{2} \text{ lbs.}$$

§ 157. **Balances.**—When a lever is employed to determine the weight of a substance it is called a balance. This may consist of a beam supported at its middle point, with pans hanging from either end, one of which holds the substance to be weighed and the other the weights; or the arms of the beam may be of different but fixed lengths and the weight may slide upon it; or the arms themselves may vary by the movement of the fulcrum.

§ 158. **The Common Balance.**—Here the arms are equal, and the beam is nicely balanced on a pivot.

A balance should be so constructed, that—

1. When the weights in the scale-pans are equal, the beam should be perfectly horizontal.

2. When the weights differ but slightly, the *deviation* of the beam from the horizontal position

Fig. 87.

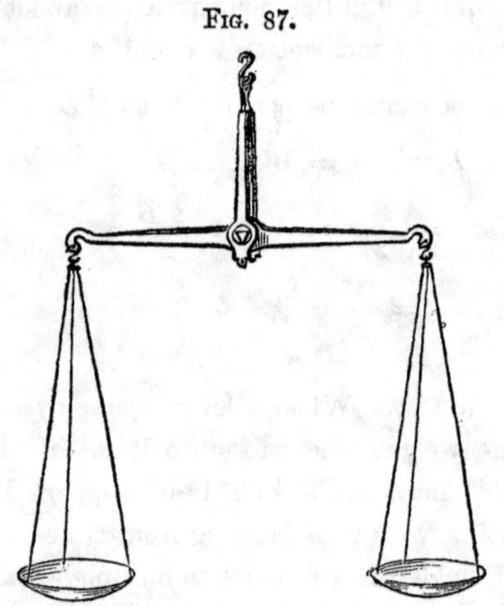

should be considerable. The *sensibility* of the balance should be delicate.

3. When disturbed the balance should quickly resume its original position. The balance should exhibit *stability*.

If a balance be true, false weights may be detected by placing them and the substance to be weighed alternately in different scale-pans; and if the weights be true, a false balance may be similarly detected.

§ 159. The *true* weight of a body can be ascertained with a *false* balance, if the weights used are correct, in the following manner :

Let W be the *real* weight of a body, and suppose it found to weigh a lbs. when placed in one scale-pan, and b lbs. when placed in the other. Let x and y be the arms of the balance, unknown. Then by the principle of moments

$$W \times x = a \times y \text{ and } W \times y = b \times x$$

$$\therefore W^2 \times xy = ab \times xy \therefore W = \sqrt{ab}$$

Or, the *true* weight is the square-root of the product of the *apparent* weights when the body is weighed in each scale-pan.

§ 160. **The Common, or Roman Steelyard.**—This is a balance with unequal arms. The body to

FIG. 88.

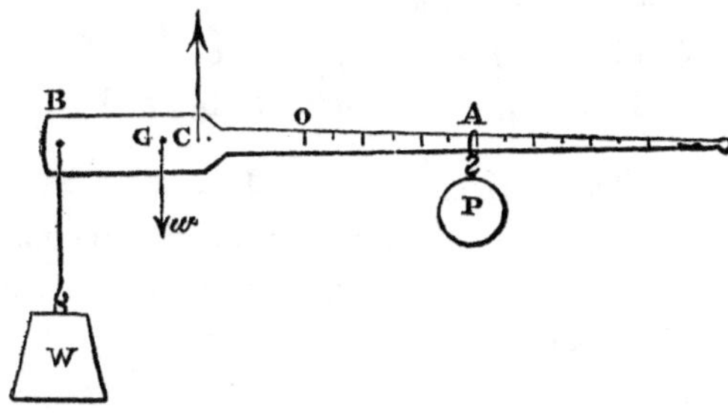

be weighed hangs from the end of one arm of the beam, and the weight employed to measure it slides on the other arm of the beam.

Let the beam be suspended at C, and let w be its weight acting at G, so that the weight P at the point O is able to keep the beam in equilibrium. Hence $w \times CG = P \times CO$.

Suppose a weight W suspended from B, and that the beam is in equilibrium when the moveable weight P is at A. Then

$$W \times BC + w \times GC = P \times AC$$
$$= P \times AO + P \times OC$$
$$\text{But } w \times GC = P \times OC$$
$$\therefore W \times BC = P \times AO$$

or

$$\frac{W}{P} = \frac{AO}{BC}.$$

Hence, the distance from the point O at which the weight P must hang varies directly with the magnitude of W.

If $OA = 1$ inch when $W = 1$ lb.
$OA = 2$ inches „ $W = 2$ lbs.
$OA = 3$ „ „ $W = 3$ lbs., and so on.

§ 161. **The Danish Steelyard.**—This instrument consists of a straight bar with a heavy knob at one end (fig. 89). The substance to be weighed is suspended at the other end, and the fulcrum O is moved so as to equate the moment of the body weighed with the moment of the weight of the beam.

Fig. 89.

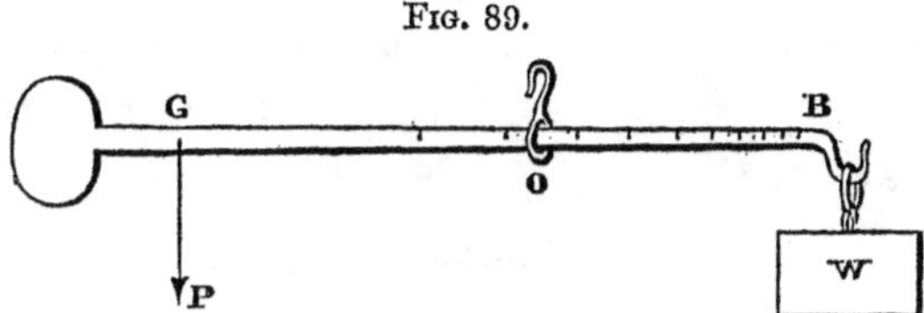

Let P equal the weight of the beam at G, and W the weight of the body suspended at B; then

$$P \times GO = W \times OB$$
$$\therefore P \times (GB - OB) = W \times OB$$
$$\therefore OB = \frac{P}{P+W} \cdot GB.$$

Hence, P and GB being fixed quantities, the beam is graduated by making W equal to 1 lb., 2 lbs., 3 lbs. successively.

Thus if $W = P$, then $OB = \frac{1}{2} GB$
" $W = 2P$ " $OB = \frac{1}{3} GB$
" $W = 3P$ " $OB = \frac{1}{4} GB$, and so on.

§ 162. Other balances are frequently formed by altering the inclination of a bent lever, and by indicating the corresponding change in the moment of the weight of the lever about its fulcrum. The common letter-weight, shown in fig. 90, is an example of this kind of balance. The fulcrum is at C, near the shorter arm of the lever. The letter to be weighed is suspended from B. When no weight hangs from B, the point at which the whole weight of the instrument acts is immediately under C, the point of suspension, and the index, which always remains vertical, points to zero. But if anything be hung from B, the balance assumes a different po-

FIG. 90.

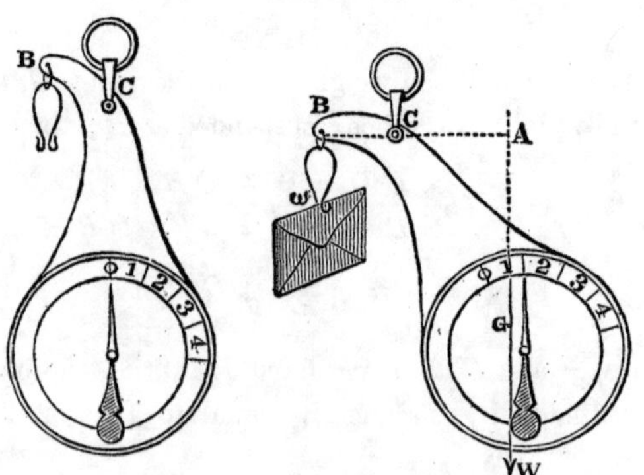

sition and the moment of its weight about C increases. If W be the weight of the instrument acting at G, and w be the weight attached to B, we have $W \times A\,C = w \times B\,C$, when equilibrium exists. If weights of $\frac{1}{2}$ oz., 1 oz., &c. be separately suspended from B, and if the corresponding positions of the vertical index be marked, these marks serve as a scale of weights.

§ 163. **General Properties of Moments.**—Since the moment of a force measures its tendency to produce rotation about a fixed point, it is evident that in order that a body, acted upon by several forces, may be in equilibrium, the various tendencies to rotation must counterbalance one another, or the sum of the positive moments with respect to any point must equal the sum of the negative moments. Hence, *when several forces act at different points of a rigid body and are in equilibrium, the algebraical sum of the moments of the forces about any point must vanish.*

§ 164. Again, since the resultant is that force which produces the same effect as its components, the rotatory tendency of the resultant is equal to the sum of the rotatory tendencies of the components. In other words, *the moment of the resultant about any point is equal to the algebraical sum of the moments of the forces about the same point.*

This proposition has already been proved in the case of parallel forces (§§ 141-2).

§ 165. Moreover, if the fixed point about which rotation is supposed to take place be in the line of action of the resultant, the moment of the resultant about this point is zero, and consequently *the algebraical sum of the moments of a number of forces about any point in the line of action of their resultant is zero.*

This proposition has also been separately proved in the case of two forces meeting at a point (§ 153).

EXERCISES.

1. Two equal and weightless rods are jointed together and form a right angle. They move freely about their common point. Find the ratio of the weights that must be suspended from their extremities that one of them may be inclined to the horizon at 60°.
2. A rod AB moves about a fixed point B. Its weight W acts at its middle point, and it is kept horizontal by a string AC that makes an angle of 45° with it. Find the tension in the string.
3. A rod 10 ins. long can turn freely about one of its ends; a weight of 4 lbs. is slung to a point 3 ins. from this end; what vertical force at the other end is required to support it?
4. If the rod, in the preceding question, be held by a string attached to the free end of the rod and inclined to it at an angle of 120°, find the tension in the string when the rod is horizontal.
5. Two forces of 3 lbs. and 4 lbs. act at the extremities of a straight lever 12 ins. long, and inclined to it at angles of 120° and 135° respectively; find the position of the fulcrum.
6. A uniform lever is bent so that its arms make an angle of 150° with each other, if the longer arm remains horizontal when a weight of 3 ozs. hangs from its extremity, and when a weight of 12 ozs. hangs from the end of the shorter arm; find the ratio of the arms.
7. Find the true weight of a body which is found to weigh 8 ozs. and 9 ozs. when placed in each of the scale-pans of a false balance.
8. In a common steelyard the weight of the beam is 10 lbs., and the distance of its centre of gravity from the fulcrum is 2 ins.; find where a weight of 4 lbs. must be placed to balance it.

9. A Danish steelyard is 36 ins. long, and its weight, 2 lbs., acts at a point 6 ins. from one end. At the other end is hung a body weighing 10 lbs.; find the position of the fulcrum.

10. If in the preceding question the fulcrum were $7\frac{1}{2}$ ins. from the thin end, what would be the weight of the body?

11. A beam 3 ft. long, the weight of which is 10 lbs., and acts at its middle point, rests on a rail, with 4 lbs. hanging from one end and 13 lbs. from the other; find the point at which the beam is supported; and if the weights at the two ends change places, what weight must be added to the lighter to preserve equilibrium?

12. Show that the sum of the moments of two forces represented by the two sides of a triangle about any point in the base is constant.

XXIII. *General Conditions of Equilibrium of Forces in one Plane—Recapitulation.*

§ 166. **Conditions of Equilibrium of any number of forces acting in one plane at different points of a rigid body.**—We are now in a position to give the results of the solution of the problem (with respect to forces in one plane) which we stated at the opening of this chapter to be the problem of Statics.

We have seen, that when any number of forces act at different points of a rigid body and in different directions, each force is capable of being resolved into two, one parallel to a fixed direction and the other perpendicular to it. The algebraical sum of all these forces in each of these two directions can be ascertained, and in order that there should be equilibrium, these two sums must separately vanish. But this condition is not, by itself, sufficient. For the algebraical sum of a number of forces will vanish if the sum of the forces in one direction is equal to the sum of the forces in the opposite direction, *i.e.*,

Fig. 91.

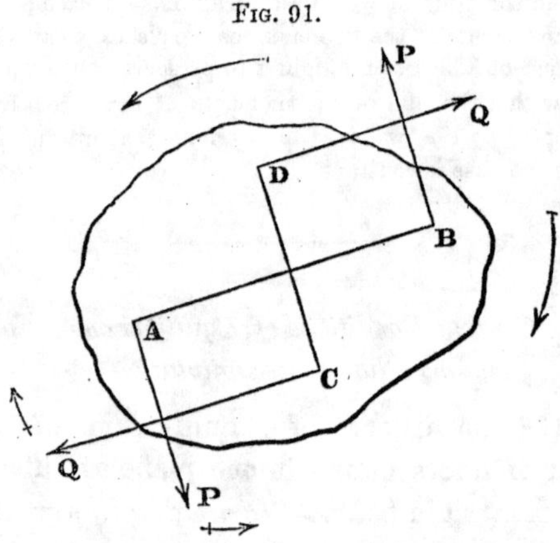

if the forces can be replaced by two *equal and opposite* resultant forces. But two equal and opposite forces will not produce equilibrium unless they act *at the same point*. Otherwise they form a couple (§ 146) and cause rotation. To produce equilibrium it is necessary that this couple should be counterbalanced by another couple, of equal moment, and acting in the opposite direction. Thus, suppose all the forces resolved in one direction are equivalent to P and P, acting at A and B, and all the forces resolved in a direction at right angles are equivalent to Q and Q, acting at C and D, the algebraical sums of the forces in each of these directions would separately vanish, and no motion of *translation* would take place. But in order that there should be no *rotation* likewise, it is necessary that the moment of P should be equal to the moment of Q; in other words, that the algebraical sum of the *moments* of the forces about any point should also vanish.

Hence the general conditions of equilibrium are three :—

$$\text{(i)}\ X_1 + X_2 + X_3 + \ldots\ldots\ldots = 0$$
$$\text{(ii)}\ Y_1 + Y_2 + Y_3 + \ldots\ldots\ldots = 0$$
$$\text{(iii)}\ P_1 p_1 + P_2 p_2 + P_3 p_3 + \ldots = 0$$

where X_1 X_2 . . . and Y_1, Y_2 . . . are the components of the force P_1, P_2 . . . in any two directions at right angles to each other, and $P_1 p_1$, $P_2 p_2$ . . . are the moments of P_1, P_2 . . . about any fixed point.

§ 167. **Recapitulation.**—In investigating the conditions of equilibrium when two or more forces act simultaneously on a body we have arrived at the following results :—

In order that there may be equilibrium when—

(i) Two forces act on a body,

(1) they must be equal in magnitude, (2) opposite in direction, and (3) act at the same point.

(ii) Three forces act on a body they must

(1) either pass through a point, and be capable of being represented by the sides of a triangle taken in order;

(2) or, be parallel, in which case the algebraical sum of the forces and of their moments about any point must vanish.

(iii) Several forces act on a body and

I. pass through a point.

They must be capable of being represented by the sides of a polygon taken in order.

II. are parallel.

The algebraical sum of the forces and their moments about any point must vanish.

III. act at different points, and in different directions.

The algebraical sums of the components of the forces in each of two directions at right angles must separately vanish; and the algebraical sum of the moments of the forces about any point in their plane must likewise vanish.

The three conditions involved in each of these cases are sufficient for the equilibrium of a body under the action of any number of forces in one plane.

§ 168. In the solution of problems condition (iii) I. very frequently enables us to obtain a geometrical representation of what is sought, although the calculation necessary to obtain a numerical solution of the problem may in many cases involve a higher knowledge of mathematics than the student is supposed to possess.

Examples.—(1) A beam the weight of which acts at its middle point hangs from a nail by two

Fig. 92.

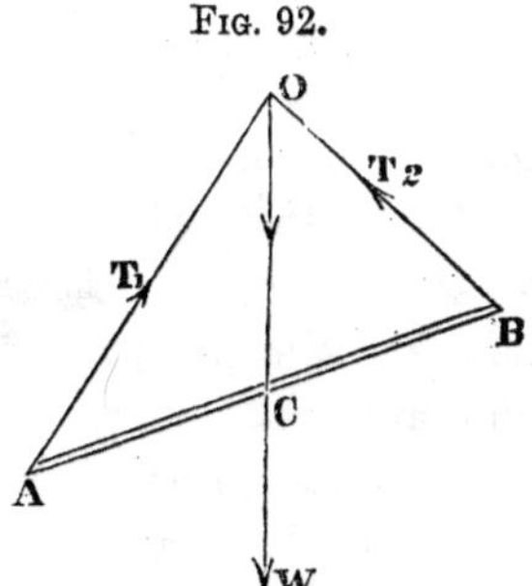

strings of given length attached to its extremities; required the position of the beam in equilibrium.

Let the weight act at C, then the forces in equilibrium are the *tensions* in the two strings and the weight; and *since these three forces must pass through a point*, the beam must so rest that the point C is vertically under the nail.

(2) A rod the weight of which acts at its middle

Fig. 93.

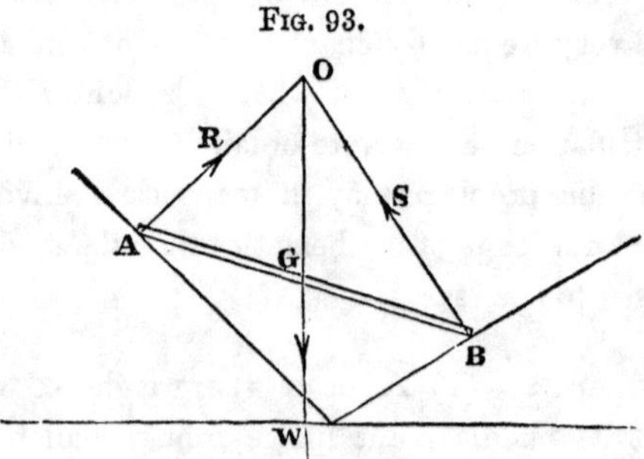

point is placed on two smooth inclined planes; required the position of the rod in equilibrium.

Here the forces acting are W at G, and R and S the reactions of the planes which are perpendicular to their surfaces. These three forces must pass through a point, and the problem of finding the position of equilibrium can be solved by a purely geometrical process, since the angles which R and S respectively make with the vertical are equal to the inclinations of the planes to the horizon.

(3) A ladder the weight of which is W and acts at a point one-third of its length from the foot, is made to rest against a smooth vertical wall, and

inclined to it at an angle of 30°, by a force applied horizontally to the foot; find the force.

Let F be the force required. Then the forces

FIG. 94.

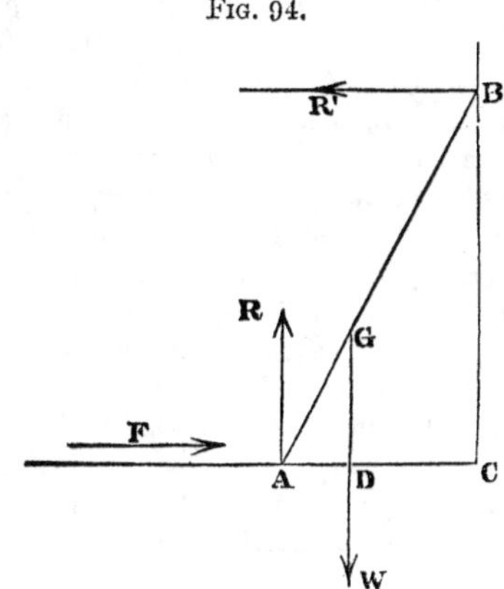

acting are W at G, the reaction of the ground R at A, and the reaction of the wall R' at B.

These form two couples which preserve equilibrium. By (iii) III. $F = R'$ and $R = W$, and, by taking moments about A, we have

$$W \times AD = R' \times CB$$

and $AD = \frac{AG}{2} = \frac{AB}{6}$; and $CB = \frac{AB\sqrt{3}}{2}$.

Also $R' = F$

$$\therefore \frac{W}{6} = F \cdot \frac{\sqrt{3}}{2}, \text{ or } F = W \frac{1}{3\sqrt{3}}.$$

EXAMINATION.

1. Show that if three forces act at a point, and produce equilibrium, they must lie in the same plane.
2. A rod *A B*, 3 ft. in length, the weight of which acts at a point 1 ft. from *A*, is hung over a smooth peg, by a cord 5 ft. in length; find the length of the cord on each side of the peg when the rod is in equilibrium.
3. Two forces of 10 lbs. and 12 lbs. act along the sides *A C*, *C B* of an equilateral triangle; find their resultant.
4. Six forces of 1 lb., 2 lbs., 3 lbs., 4 lbs., 5 lbs., and 6 lbs., acting at a point, make equal angles with one another; find their resultant.
5. Prove that when two forces act at a point, their resultant is nearer to the greater force.
6. Apply the above to the case of parallel forces.
7. Give the conditions of equilibrium where two forces act on a body.
8. A weight of 10 lbs. is placed on a square board 5 ins. from one side, and 10 ins. from the opposite side, along the line that bisects its width. The board is suspended at its corners by four vertical strings; find the tension in each string.
9. Explain the proposition known as the *polygon of forces*.
10. Two equal forces of 8 lbs. act at an angle of 120°; a third force of 6 lbs. acts at the same point, and at right angles to their plane; find the resultant of the three forces.
11. A lever with a fulcrum at one end is 3 ft. in length. A weight of 28 lbs. is suspended from the other end. If the weight of the lever is 2 lbs. at its middle point, at what distance from the fulcrum will an upward force of 50 lbs. preserve equilibrium?
12. Two forces of 4 lbs. and 8 lbs. act at the end of a bar

18 ins. long and make angles of 120° and 90° with it; find the point in the bar at which the resultant acts.

13. Assuming the parallelogram of forces, show how the fulcrum of a lever of any shape may be determined.

14. What is the condition that two *couples* may produce equilibrium?

15. A picture the weight of which is 4 lbs. is suspended from a nail by means of a flexible cord. The top of the picture is horizontal, and the angle between the two parts of the cord is 30°; find the tension in the cord.

University of London Examination Questions.

I. MATRICULATION.

16. State and explain the proposition known as the 'Parallelogram of Forces.'—Jan. 1869.

17. A weight of 24 lbs. is suspended by two flexible strings, one of which is horizontal, and the other is inclined at an angle of 30° to the vertical direction. What is the tension in each string?—June 1869.

18. If two forces acting on a point are represented in magnitude and direction by two sides of a triangle, under what circumstances will the third side correctly represent their resultant? Forces of 20 and 10 act along the sides $A B$ and $B C$ respectively of an equilateral triangle; find the magnitude of their resultant.—June 1869.

19. Show how to find the resultant of three given forces acting on a point; and prove that to produce equilibrium their directions must lie in the same plane.—Jan. 1870.

20. Find the ratio of the power to the weight for equilibrium on a bent lever of the first kind, when the forces act

at right angles to the arms. Supposing the arms make an angle of 120° with each other, and have the relative lengths 1 and 5, find the magnitude and point of application of the resultant of the power and weight when the lever is in equilibrium.—Jan. 1870.

21. What is meant by the moment of a force about a given point? How is its magnitude determined?—June 1870.

22. If several forces which do not balance act in the same plane upon different points of a solid body one point of which is fixed, what condition must be fulfilled in order that the body may be in equilibrium? Show that this condition cannot be fulfilled unless the fixed point is in the plane of the forces.—June 1870.

23. Show that as the angle between two forces is increased their resultant is diminished.—Jan. 1871.

24. Two forces of the magnitude 5 and 11 act at angles of 60°, 90°, 120° respectively; compare their resultants in the three cases.—Jan. 1871.

25. A weight of 56 lbs. is attached to a straight lever without weight, at a distance of 3 ins. from the fulcrum, and is balanced in one case by a power of 6 lbs., and in another case by a power of 16 lbs.; find in each case the pressure on the fulcrum, and also the distance between the points of application of the power and weight, when they are applied.

(*a*). On the same side of the fulcrum.

(*b*). On opposite sides of the fulcrum.

Show how the answer to each part of the question would be affected if the lever weighed 9 lbs., and its centre of gravity were at the fulcrum.—Jan. 1871.

26. Assuming the principle of the Parallelogram of Forces, show what must be the relation, in order that there may be equilibrium on the inclined plane, between the power, the weight, and the pressure against the plane.—June 1871.

27. A pole of 12 ft. long, weighing 25 lbs., rests with one end against the foot of a wall, and from a point 2 ft. from the other end a cord runs horizontally to a point in the wall 8 ft. from the ground; find the tension of the cord and the pressure of the lower end of the pole.—June 1871.

28. A solid roller, with an axle projecting from one end, is suspended horizontally by two vertical cords—one of them attached to the end of the roller opposite to the axle, the other to the middle of the axle; the roller is 4 ft. long, and weighs 27 lbs.; the axle is 1 ft. long, and weighs 1 lb. Find the weight supported by each cord.—June 1871.

29. Show *why* a lever is in equilibrium when the power and weight are to each other inversely as the perpendicular distances of their lines of action from the fulcrum.—Jan. 1872.

30. Show how to resolve a given force into two components, one of which has a given magnitude and acts parallel to a given straight line.

As a special case, resolve a force of magnitude 12, acting horizontally from left to right, into two components, one of which is a force of magnitude 25 acting vertically upwards.—Jan. 1872.

31. A man wheels a loaded wheelbarrow along a level road; point out the conditions which determine how much of the total weight of load and barrow is supported by the wheel, and how much is supported by the man.—Jan. 1872.

32. Three equal forces act in one plane on a point in such a way that each of them makes an angle of 120° with each of the other two; prove that the forces will balance.—June 1872.

33. Employ the above proposition to show that the resultant of the forces 7 and 14 acting at an angle of 120° is the same as the resultant of forces 7 and 7 acting at an angle of 60°.—June 1872.

34. Show how it is possible for a sailing vessel to make way in a direction different from that of the wind. Why cannot a round tub be steered at as great an angle to the direction of the wind as a long-boat?—June 1872.

35. When a horse is employed to tow a barge along a canal the tow-rope is usually of considerable length; give a definite reason for using a long rope instead of a short one. Show whether the same considerations hold good in relation to the length of the rope when a steam-tug is used instead of a horse.—June 1872.

36. If a man wants to help a cart up hill is there any mechanical reason why he should put his shoulder to the wheel, instead of pushing at the body of the cart? And if so, show at what part of the wheel force can be applied with the greatest effect.—Jan. 1873.

37. In a combination of one fixed and one moveable pulley, where the power acts horizontally, and the fixed end of the string is attached to the upper block, find the direction of the total resultant pressure on the upper block, and its magnitude as compared with the power. —Jan. 1873.

38. A body weighing 6 lbs. is placed on a smooth plane, which is inclined at 30° to the horizon; find the two directions in which a force equal to the body may act to produce equilibrium. Also find what is the pressure on the plane in each case.—Jan. 1873.

39. Two men are carrying a block of iron, weighing 176 lbs., suspended from a uniform pole 14 ft. long; each man's shoulder is 1 ft. 6 in. from his end of the pole. At what point of the pole must the heavy weight be suspended, in order that one of the men may bear $\frac{4}{5}$ of the weight borne by the other?—June 1873.

40. A heavy plummet is immersed in a stream, the string being held by a person standing on the bank. The string is found to settle in a sloping position. Show by means of a sketch the three forces which keep the plummet in equilibrium.—June 1873.

41. State the Parallelogram of Forces. Explain the meaning of the terms employed in your statement; apply it to show that if four forces acting on a point be represented by the sides of a rectangle taken in order, they will be in equilibrium.—Jan. 1874.

42. A substance is weighed from both arms of an unequal balance, and its apparent weights are 9 lbs. and 4 lbs.; find the ratio between the arms.—Jan. 1874.

43. If in the third system of pulleys (Fig. 40) the pulleys are $1\frac{3}{4}$ in. in diameter, and the strings are parallel and attached at given points *A*, *B*, and *C* to the rod supporting the weight, to what point of the rod should the weight *W* be attached, so that the horizontal direction of the rod may be maintained?—Jan. 1874.

44. Explain the meaning of the word Composition and Resolution of Forces, and show how forces may be compounded and resolved. A point is acted upon by a force whose magnitude is unknown, but whose direction makes an angle of 60° with the horizon. The horizontal component of the force is known to be 1·35. Determine the total force, and also its *vertical* component.—June 1874.

45. Assuming the truth of the Parallelogram of Forces, show how to find the position and magnitude of the resultant of two parallel forces, *P* and *Q*, acting at different points of a rigid body.—June 1874.

46. Two uniform heavy rods, *A C*, *B C*, rigidly connected

FIG. 95.

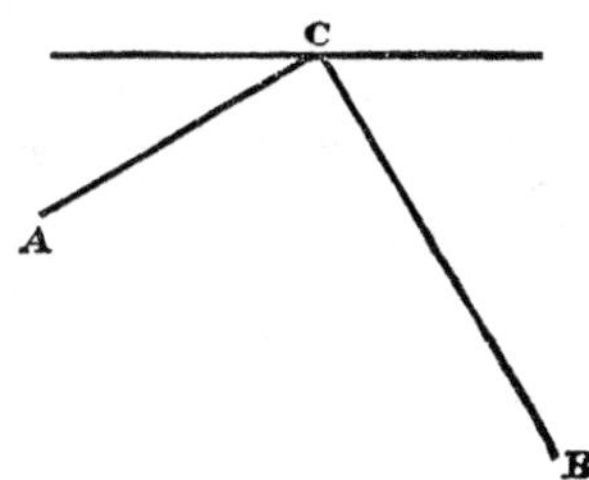

together, are capable of turning round a horizontal axis at C; find the mechanical conditions which determine the position of equilibrium.—June 1874.

II. PRELIMINARY SCIENTIFIC 1st M. B.

47. To each end of a uniform straight rod 100 ins. long, and weighing 12 lbs., is fastened one end of a flexible string 140 ins. long, to which a weight of 9 lbs. is attached at a point 60 ins. from one end. In what position will the rod remain in equilibrium about a pivot through the middle? and where must the pivot be placed in order that the rod may be balanced when horizontal?—1869.

48. The extremities of the horizontal diameter of a circular disc, weighing 6 ozs., are nailed against a wall, and to a point in the edge of the disc at $\frac{1}{12}$ of the whole circumference from one of the nails a weight of 4 ozs. is attached; find the pressure upon each nail.—1870.

49. A rod AB, 5 ft. long, without weight, is hung from a point C by two strings which are attached to its ends and to the point; the string AC is 3 ft., and BC is 4 ft. in length, and a weight of 2 lbs. is hung from A, and a weight of 3 lbs. from B; find the tensions of the strings and the condition of equilibrium.—1871.

50. Prove that the sum of the moments of any two parallel forces about a point is equal to the moment of their resultant about the same point.—1871.

51. The weight of a window-sash 3 ft. wide is 5 lbs., each of the weights attached to the cords is 2 lbs.; if one of the cords be broken, find at what distance from the middle of the sash the hand must be placed to raise it with the least effort.—1872.

52. The moments of two forces about any point in the line of action of their resultant are equal and opposite. Deduce from this principle a rule for finding the line

of action of the resultant of two given parallel forces; and discuss the result of applying the rule to the case of equal parallel forces acting in opposite directions.—1872.

53. Show how to graduate the Roman steelyard, taking account of the weight of the instrument.—1873.

54. A light smooth stick 3 ft. long is loaded at one end with 8 ozs. of lead; the other end rests against a smooth vertical wall, and across a nail which is 1 ft. from the wall. Find the position of equilibrium and the pressure on the nail and on the wall.—1873.

55. Show that if two forces act upon a body the moment of the one about a point in their resultant is equal to that of the other about the same point.—1874.

CHAPTER VIII.

CENTRE OF GRAVITY.

XXIV. *Properties of Centre of Gravity—Equilibrium of a body on a hard surface.*

§ 169. The weight of a body is the force with which a body, free to fall, is urged towards the earth's centre; the point at which this force acts is called the centre of weight, or *centre of gravity.* Hence the centre of gravity may be defined as *the point at which the whole weight of a body may be supposed to act.*

If the body be supported at this point the body will rest in any position whatever.

We shall see later that such a point really exists; at present we shall assume that there is a point *in* or *with respect to* every body, at which the whole weight acts.

§ 170. **If a body be suspended at any point the Centre of Gravity will be in the vertical line drawn through the point of suspension.**—Let the body be suspended at S. Let W be the weight of the body and G its centre of gravity.

Then W acts at G vertically downwards. The reaction of the point of support is equivalent to a

FIG. 96.

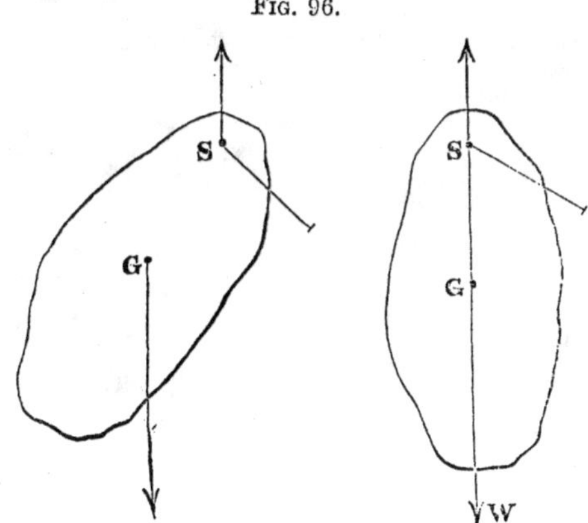

force acting vertically upwards at S. For equilibrium these two forces must pass through the same point (§ 167, 1). Therefore G must be in a vertical line with S. In any other position of the body the moment of W about S would cause rotation.

§ 171. **Method of finding the Centre of Gravity of uniform Laminæ.**—The foregoing property enables us experimentally to find the centre of gravity of uniform laminæ or flat bodies of inconsiderable thickness. For if the figure be first suspended at one point, and then at another point, the centre of gravity will be in the verticals through each of these points of suspension; and if these vertical lines be marked on the body their point of intersection will be the centre of gravity required.

If the centre of gravity be known, we can deter-

Fig. 97.

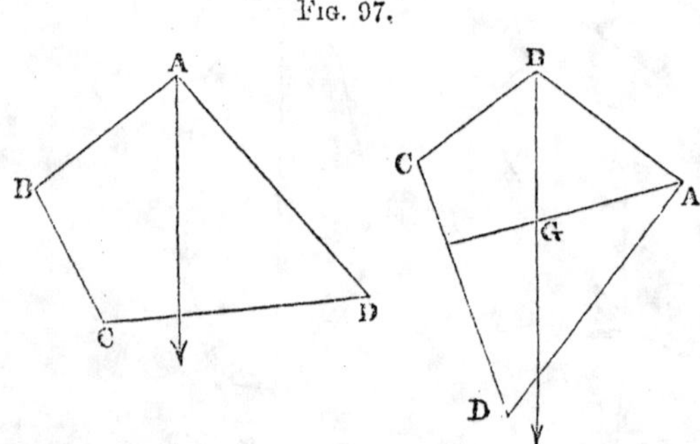

mine the point in the opposite side of the surface which will be vertically under the point of suspension. For if the line joining the centre of gravity with the point of suspension be produced to cut the opposite side, the point of intersection will be the point required.

§ 172. If a body rest on a hard plane surface, it will stand or fall according as the vertical from the Centre of Gravity falls within or without the base.—The first condition that a body may rest on a surface is that the surface shall be strong enough to support the body. Otherwise the body obeys its natural impulse and falls through. If the material be sufficiently strong it exerts a force perpendicular to its surface, which may be considered as acting at some point in the surface in contact with the body. If the vertical through *G* (fig. 98) fall within the base, the reaction of the plane will be in the same line with

Fig. 98.

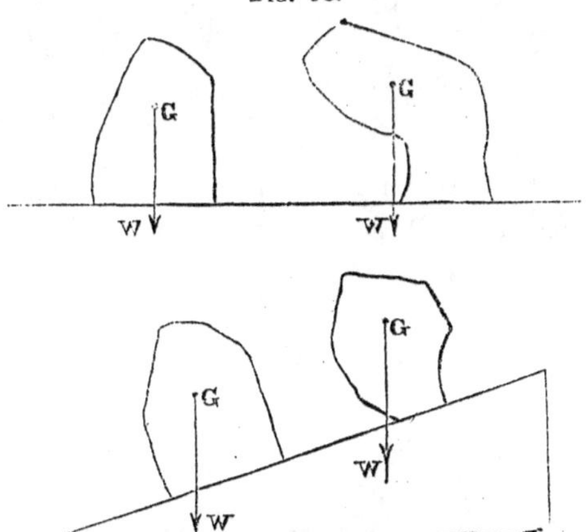

the direction of *W* and there will be equilibrium. But if the vertical through *G* fall beyond the base, *W* and the reaction cannot be in equilibrium and the body will fall.

The condition is the same whether a body rests on a horizontal or inclined plane.

§ 173. When it is said that the vertical from the centre of gravity must fall within the *base*, it must be observed that by the base of a body is meant the line drawn round the points of support. For if a

FIG. 99.

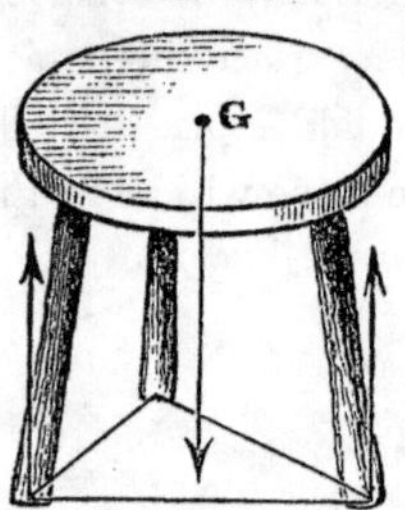

body rest on three points of support, as in fig. 99, these three reactions will have a resultant which in equili-

FIG. 100.

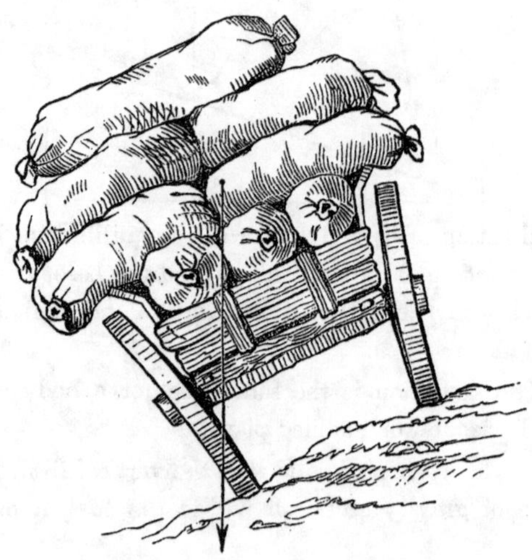

brium passes through G. If the body rests on more points of support the same holds good. If, however, the vertical from G (fig. 100) falls *without* the circumference of the points of support, the resultant of the several reactions will not pass through G and equilibrium will not exist.

§ 174. **Stable, unstable, and neutral Equilibrium.**—Equilibrium may be of *three* kinds:

1. The body may be in such a position that if

Fig. 101.

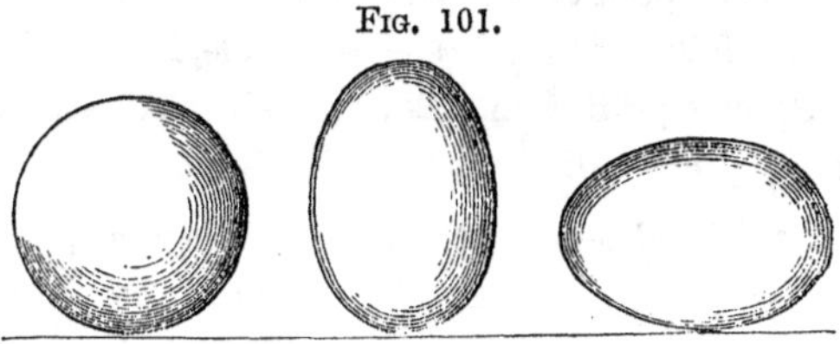

slightly displaced it tends to *return to* its original position, in which case the equilibrium is ***stable.***

Fig. 102.

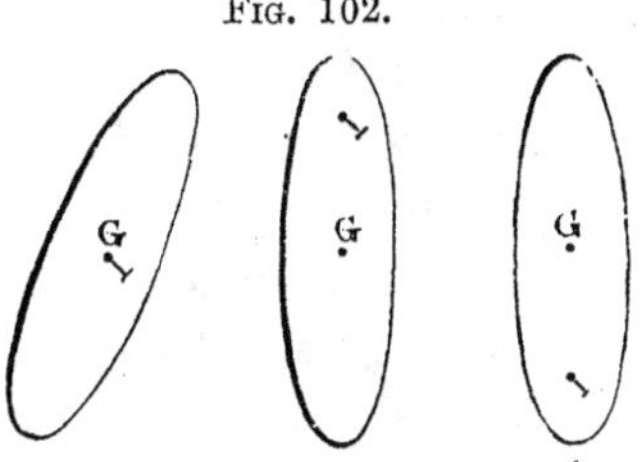

2. Or, it may tend to *move further away from* its position, in which case the equilibrium is ***unstable.***

3. Or, it may *remain in* its new position, in which case the equilibrium is *neutral.*

These conditions are illustrated when a body rests on a hard surface or when suspended by a smooth peg. An egg on either end is in unstable equilibrium, when resting on a longer side it is in stable equilibrium. A sphere or cylindrical roller resting on a horizontal surface is in neutral equilibrium. A disc suspended at its centre of gravity will rest in neutral equilibrium. If suspended at any other point there are two positions *only* in which it will rest. In the one it fulfils the conditions of stable, in the other of unstable equilibrium. These positions are shown in figs. 101 and 102

It will be observed that a body when displaced assumes a position of unstable equilibrium in passing

FIG. 103.

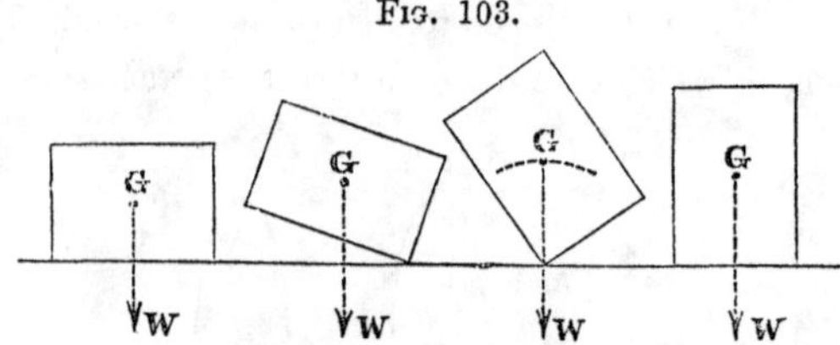

from one position of stable equilibrium into another; and that bodies having several surfaces on which they can rest have positions of stable and unstable equilibrium corresponding to each. Hence there are some positions in which the equilibrium is more stable than in others.

§ 175. **Position of Centre of Gravity in stable and unstable Equilibrium.**—When a body is in stable equilibrium it will, if slightly displaced, return to its original position, and the centre of gravity of the body will describe an arc, first rising and then falling. If the body be further displaced the centre of gravity will gradually rise till it has reached its highest position, in which case the equilibrium will be unstable.

Hence, in stable equilibrium the centre of gravity occupies the lowest possible position; and in un-

Fig. 104.

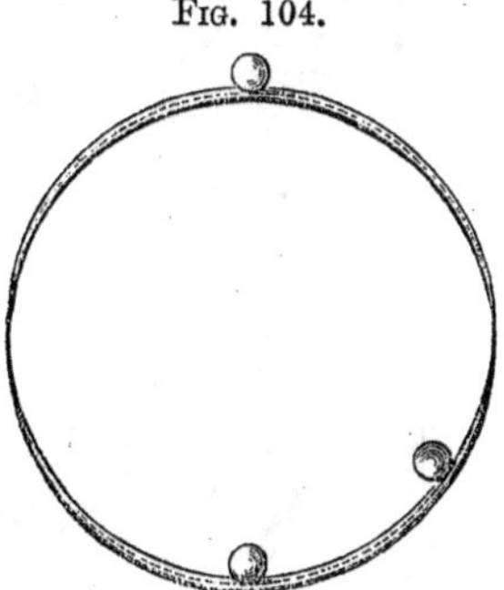

stable it occupies its highest position; and therefore when a body is at rest its centre of gravity must occupy its highest or lowest position. This may be illustrated by taking a wooden hoop (fig. 104) on the inside and outside of which a groove is cut capable of holding a marble. If the hoop be stood in a vertical position it will readily be seen that there are two positions, and two only, in which the marble can rest. It may, with very great care, be made to rest on the summit, and the slightest displacement will cause it to roll down; or it may remain in stable equilibrium at the bottom of the hoop. In any other position equilibrium is impossible.

§ 176. A body seemingly in unstable equilibrium may really be in stable equilibrium, if by the addition of weights the centre of gravity is brought under the

Fig. 105.

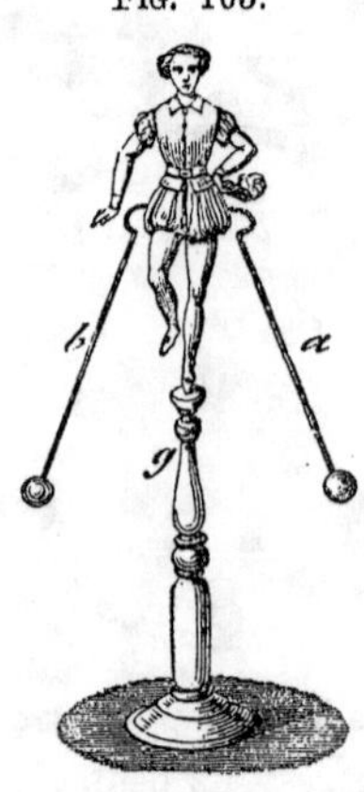

point of support. Thus, a half-crown piece will not rest with its edge on the rim of a wine-glass; but if two silver forks hang by their prongs to the coin it will be found to be supported in stable equilibrium.

In the toy represented in the annexed diagram the centre of gravity g is under the point of support, and the figure consequently can undergo a certain amount of displacement without the equilibrium being destroyed.

§ 177. **The energy of a body in its three states of Equilibrium.**—Bearing in mind the distinction between kinetic and potential energy (§ 74), we shall readily see that the potential energy or energy of position, which a body possesses, varies with its condition of equilibrium. In neutral equilibrium, the potential energy is the same for all positions of the body. In stable equilibrium the potential energy is a minimum; in other words, the body is in the most unfavourable position for doing work; whilst in unstable equilibrium the potential energy is a maximum, and the application of the smallest force can, at once, convert this energy of position into energy of motion. A body in unstable equilibrium may be said, therefore, so far as its position is concerned, to be charged with the greatest amount of potential energy it can possess.

§ 178. A body, which is prevented from sliding, is on the point of overturning when the vertical from its centre of gravity falls at the extremity of its base, in which case the body is in a position of unstable equilibrium. If a body rest on a rough plane, and the plane be inclined through such an angle that the vertical from the centre of gravity falls at the edge of its base, any further elevation of the plane will cause the body to overturn.

The inclination of the plane at which this takes place can be easily ascertained. Let G be the centre of gravity of a body resting on the inclined plane $C D$,

Fig. 106.

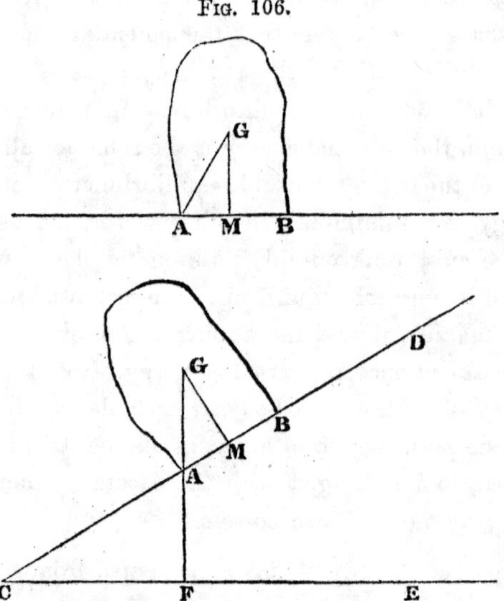

in unstable equilibrium. Then if $A B$ be the base of the body, the vertical from G passes through A, and is at right angles to the horizontal line $C E$ drawn through C. If $G M$ be drawn perpendicular to $A B$, the right-angled triangles $A G M$, $A C F$

will have the angle $G A M$ in the one equal to the angle $C A F$ in the other, and consequently the remaining angle, $A G M$, equal to the angle $A C F$. But $A C F$ is the angle of the plane, and therefore the angle at which the plane may be inclined must be less than the angle contained by the lines $G A$ and $G M$, in order that the body may be in equilibrium. Or, the body will be on the point of overturning when the angle of the plane is such that

$$\frac{\text{the height of the plane}}{\text{the base of the plane}} = \frac{AM}{GM}.$$

XXV. *Methods of finding the Centre of Gravity of Particles and Bodies.*

Different methods have to be adopted for finding the centre of gravity of bodies according to their form.

We commence with a system of heavy particles.

§ 179. **To find the Centre of Gravity of two heavy particles.**—The weight of each particle A and B may be considered as a vertical force, and the centre of gravity of the particles is the point where the resultant of these two forces acts. In consequence of the great distance of the earth's centre from the particles A and B, these two forces may be considered as parallel; and, for the same reason, the weights of the particles of all bodies may be regarded as parallel forces.

Join $A B$, and if W_1 and W_2 be the weights of the two particles, divide $A B$ into $W_1 + W_2$ parts,

Fig. 107.

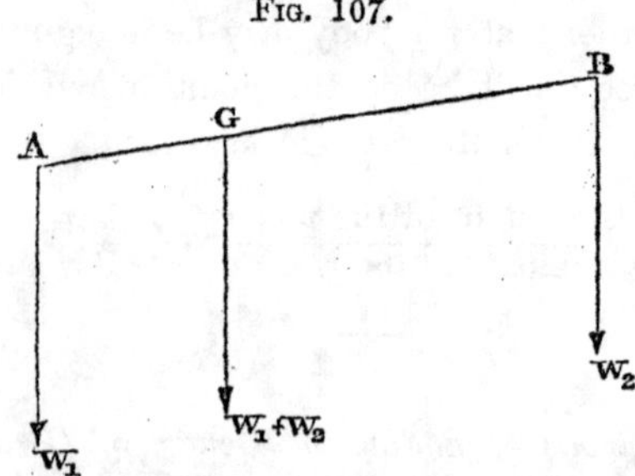

and mark off $A G$ equal to W_2 of these parts, so that $\frac{A G}{B G} = \frac{W_2}{W_1}$; then G is the point where $W_1 + W_2$ acts, *i.e.*, G is the centre of gravity of the particles.

§ 180. **To find the Centre of Gravity of a number of heavy particles.**—Since the weights of the several particles may be regarded as parallel forces, and since the centre of gravity is the point where the weight of all the particles may be supposed to act, the centre of gravity of the particles coincides with the point where the resultant of all these forces acts, and may be found in the same way as the centre of a system of parallel forces (§ 140).

§ 181. Every material body consists of particles which may be considered as so many vertical forces, and the centre of these forces is the point where their sum, which is equal to the whole weight of the body, acts. But the point where the whole weight of the body acts is the centre of gravity of the body, and therefore the centre of gravity of a body is the point at which acts the resultant of all the forces, caused by the earth's attraction upon the particles, of which the body is composed.

The centre of gravity of a body is very often thus defined; but it is preferable to start with a simpler definition, and one which does not involve the word Resultant and the theory of parallel forces.

§ 182. We have seen that the position of the centre of a system of parallel forces is independent of the direction of the forces, and the centre of gravity of a body is therefore a fixed point independent of the position of the body, and even of the direction of the force of gravitation.

§ 183. **Centre of Gravity of Homogeneous and Symmetrical Bodies.**—In finding the centre of gravity of bodies we shall consider the matter of which they are composed to be distributed uniformly, so that the weights of portions of a body will be proportional to their volumes. When we speak of the centre of gravity of a line, we suppose the line to be a thin heavy rod of uniform section; and when we consider the centre of gravity of a surface we suppose the surface to be a thin lamina, the thickness of which, being uniform, can be neglected. Hence, for the weights of thin laminæ we may substitute the areas of thin surfaces.

If a body be symmetrical about a plane, or if a surface be symmetrical about a line, the centre of gravity is in that plane or line; and the centre of gravity coincides with the geometrical centre of all figures that have such a point. Hence—

1. The centre of gravity of a straight line is its middle point.
2. The centre of gravity of a circle or of its circumference, or of a sphere or of its surface, is its centre.
3. The centre of gravity of a parallelogram or of its perimeter is the point in which the diagonals intersect.
4. The centre of gravity of a cylinder or of its surface is the middle of its axis.

§ 184. **To find the Centre of Gravity of the Surface of a Triangle.**—Let $A B C$ be the triangle. Draw the *median* line $A D$ from the point A to the centre of $B C$, and divide it into a certain number of equal parts. Through the points of division D', D'', D''' . . . draw parallels to the base $B C$, and through the points B', C', B'', C'' . . . where the parallels cut the other sides, draw lines parallel to $A\,D$. We thus form a series of parallelograms inscribed in the triangle. The line $A\,D$ passes through the centres O, O', O'' . . . of all these parallelograms, since it bisects their opposite sides.

FIG. 108.

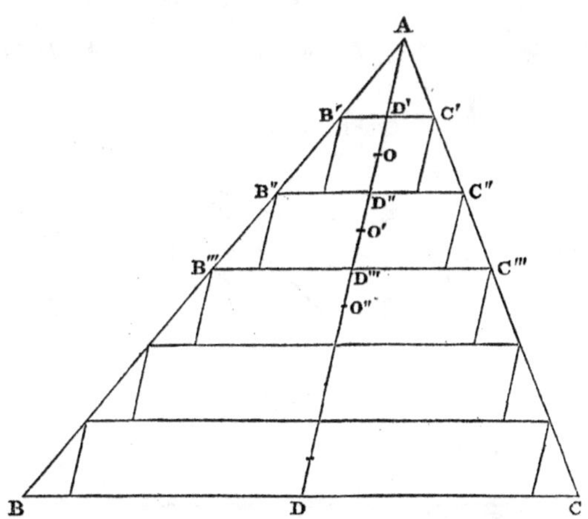

If we consider the weights of these parallelograms as parallel forces acting at the points O, O', O'' . . . the resultant of all these forces will be in the line $O\,D$. The sum of the parallelograms is less than the surface of the triangle, but it can be shown, by reasoning similar to that employed in § 26, to approach as near to the surface as ever we please, by increasing the number of divisions. The centre of gravity of all the parallelograms being situated in *A D*, it follows that the centre of gravity of the triangle is also in *A D*.

In the same way it may be shown that the centre of gravity of the triangle is in the median *C E*, (fig. 109). Hence the centre of gravity of the triangle is at *G*, where the two medians intersect.

Produce *A D* to *F*, making *D F* equal to *D G*.

Fig. 109.

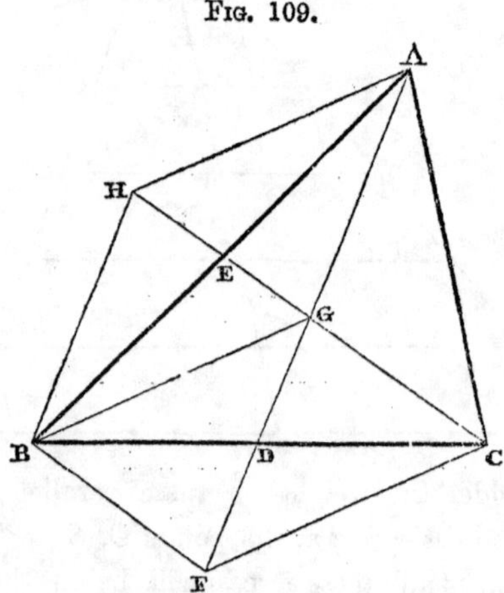

Produce *G E* to *H*, making *E H* equal to *G E*. Join *B G*, *B F*, and *F C*; join *B H* and *H A*.

Then, since the lines *B C* and *G F* are bisected at *D*, the figure *B G C F* is a parallelogram, and *B F* is parallel to *C G*. For the same reason the figure

$H A G B$ is a parallelogram, and $H B$ is parallel and equal to $A G$. Since $H B$ is parallel to $A F$, or $A G$ produced, and $B F$ is parallel to $H G$, the figure $H F$ is a parallelogram

$$\therefore H B = G F.$$

But $$H B = A G \therefore A G = G F = 2 G D$$

$$\therefore G D = \tfrac{1}{3} A D.$$

Or, the centre of gravity of the surface of a triangle is situated on a median line, and at a distance of $\frac{1}{3}$ of its length from the base.

§ 185. **Example.**—If weights, each equal to P, be placed at the angular points A, B, and C, the centre of gravity of these weights coincides with the centre of gravity of the triangle.

The resultant of the two forces at B and C is a force equal to $2 P$ at D, and this resultant, combined with P at A, gives a force of $3 P$ at G, such that $A G$ equals $2 D G$.

§ 186. **To find the Centre of Gravity of the Perimeter of a Triangle.**—We suppose the triangle to be contained by three heavy lines joined together.

The middle points A', B', C' are the centres of gravity of the three lines (§ 183, 1). Join $B' C'$ and divide it in D, so that

$$\frac{A B}{A C} = \frac{B' D}{C' D}.$$

Then D is the centre of gravity of the two sides $A B, A C$.

Join $D A'$, and divide $D A'$ in G, so that

$$\frac{D G}{G A'} = \frac{B C}{A B + A C}.$$

FIG. 110.

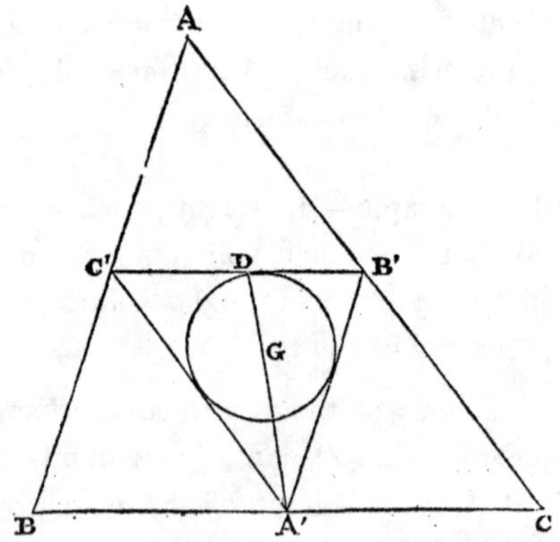

Then G is the point where the weight of the three sides acts, *i.e.*, G is the centre of gravity of the perimeter.

Since $\frac{D B'}{D C'} = \frac{A B}{A C}$ and $A' B' = \frac{1}{2} A B$; and $A' C' = \frac{1}{2} A C \therefore \frac{D B'}{D C'} = \frac{A' B'}{A' C'}$ and, therefore, the line $A' D$ is known to bisect the angle $B' A' C'$. In the same way it can be shown that G is situated on a line that bisects the angle $C' B' A'$. Hence the centre of gravity of the triangle is situated at the centre of the circle inscribed in the triangle $A' B' C'$.

§ 187. **To find the Centre of Gravity of any rectilinear figure.**—When a rectilinear figure can easily be divided into triangles, its centre of gravity can very frequently be found by finding the centre of

Fig. 111.

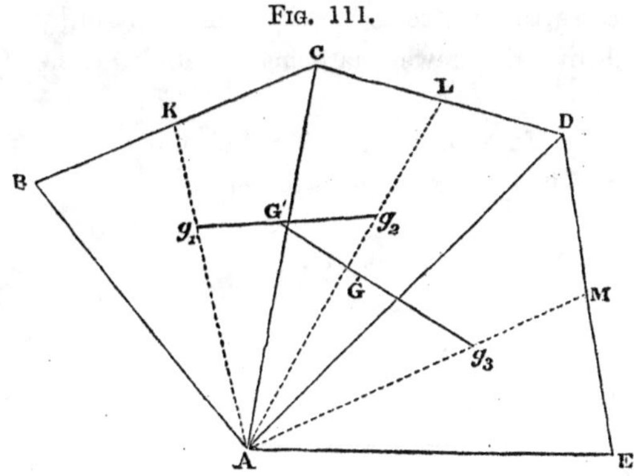

gravity of each of the triangles, and compounding the forces acting at these points by the method given in § 140.

Thus, suppose $A\ B\ C\ D\ E$ a rectilinear figure. Divide it into three triangles by lines drawn through A. Draw the median lines $A\ K$, $A\ L$, $A\ M$. Then the centre of gravity of triangle $A\ B\ C$ is at g_1, of triangle $A\ C\ D$ at g_2, and of triangle $A\ D\ E$ at g_3. Join g_1, g_2 and divide the line in G' so that

$$\frac{g_1 G'}{G' g_2} = \frac{\Delta A C D}{\Delta A B C.}$$

Join $G' g_3$ and divide this line in G, so that

$$\frac{G' G}{G g_3} = \frac{\Delta A D E}{\Delta A C D + \Delta A B C.}$$

Then G is the centre of gravity required.

This method is theoretically possible; but in many cases the centre of gravity can be found by a method involving fewer mathematical difficulties.

§ 188. **Examples.**—(1) To find the centre of gravity of half a regular hexagon.—Let $D A B C$ be

Fig. 112.

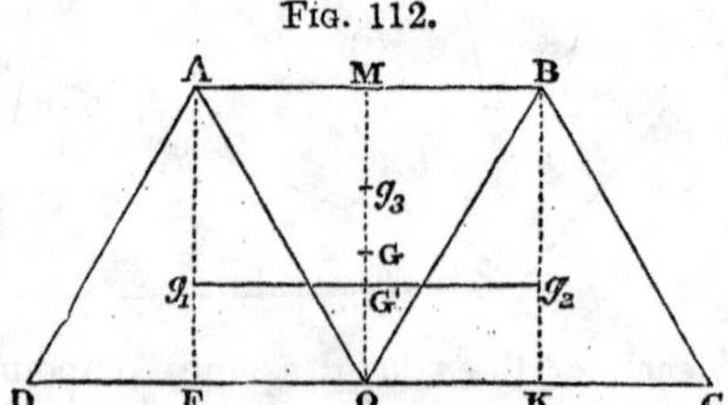

half a regular hexagon. Then the sides $D A$, $A B$, $B C$ are all equal, and if O bisects $D C$, and $O A$ and $O B$ are joined, the figure is divided into three equilateral triangles. Draw the median lines $A F$, $O M$, $B K$. These are parallel, and perpendicular to $D C$ or $A B$.

The centre of gravity of $\Delta A O D$ is at g_1 where $A g_1 = 2 F g_1$; the centre of gravity of $\Delta B O C$ is at g_2 where $B g_2 = 2 K g_2$.

Join $g_1\, g_2$. Then, since these triangles are equal, their centre of gravity is at a point G' which bisects the line $g_1\, g_2$. This point, therefore, is on the line $O\,M$, and $O\,G' = \frac{1}{3}\,O\,M$. Now, the centre of gravity of the $\Delta\, A\, O\, B$ is at g_3 where $M\, g_3 = \frac{1}{3}\,O\,M$, and therefore $M\,g_3 = O\,G' = G'\,g_3$. Divide $G'\,g_3$ in G so that

$$\frac{G'\,G}{G\,g_3} = \frac{\Delta\,A\,O\,B}{\Delta\,A\,D\,O + \Delta\,B\,O\,C} = \frac{1}{2}.$$

Then G is the centre of gravity required, and $O\,G = O\,G' + G'\,G$ or $O\,G = \frac{4}{9}\,O\,M$.

(2) To find the centre of gravity of a figure

FIG. 113.

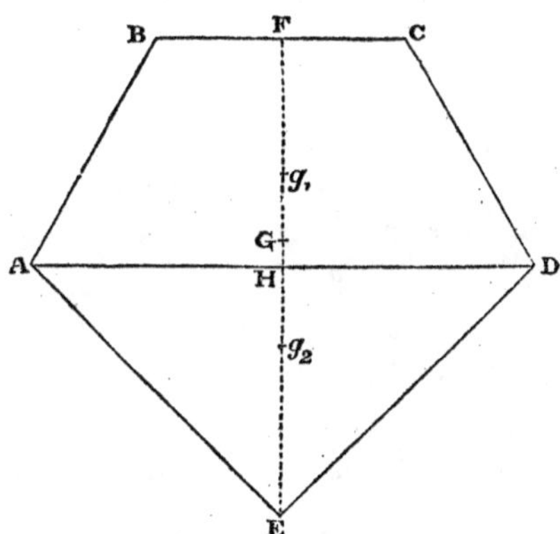

which consists of half a regular hexagon, having an isosceles right-angled triangle on its longest side.

Let $A B C D E$ be the figure which is symmetrical about the line $E F$. The centre of gravity is therefore somewhere in the line $E F$.

Let $A B = a = B C = C D$.

Then $A D = 2 a$ and $H E = a$.

Hence the area of the half-hexagon is $3 \frac{a \,.\, F H}{2}$ $= \frac{3 a^2 \sqrt{3}}{4}$.

The area of triangle $A D E$ equals $\frac{A D \times H E}{2}$ $= a^2$.

Let g_1 be the centre of gravity of half-hexagon, then $H g_1 = \frac{4}{9} F H$.

Let g_2 be the centre of gravity of triangle, then $H g_2 = \frac{1}{3} H E$.

Divide $g_1 g_2$ in G, so that

$$\frac{g_1 G}{G g_2} = \frac{a^2}{\frac{3 a^2 \sqrt{3}}{4}} = \frac{4}{3 \sqrt{3}}.$$

Then G is the centre of gravity required. Let $g_1 g_2 = b$ and $G g_1 = x$,

then $$\frac{x}{b - x} = \frac{4}{3 \sqrt{3}} \therefore x = \frac{4 b}{4 + 3 \sqrt{3}},$$

where $$b = \frac{4}{9} F H + \frac{1}{3} H E = \left(\frac{4}{9} \frac{\sqrt{3}}{2} + \frac{1}{3}\right) a =$$ $\frac{2 \sqrt{3} + 3}{9} \,.\, a$.

Hence $G g_1 = \frac{4}{9} \cdot \frac{2}{3} \frac{\sqrt{3} + 3}{\sqrt{3} + 4} \cdot a$, which gives the distance of the centre of gravity from g_1.

To find the distance of G from H, we can subtract $G g_1$ from $H g_1$ which gives

$$H G = \frac{2}{33} (3 \sqrt{3} - 4) a.$$

We can also take moments about H. Then, since the moment of the resultant equals the sum of the moments of the forces, we have area $A B C D \times H g_1$ − area $A E D \times H g_2$ = whole area $\times H G$.

$$\frac{3 a^2 \sqrt{3}}{4} \times \frac{4}{9} \frac{\sqrt{3}}{2} a - a^2 \times \frac{1}{3} a = \left(\frac{3 a^2 \sqrt{3}}{4} + a^2\right) G H$$

$$\therefore H G = \frac{2}{33} (3 \sqrt{3} - 4) a.$$

The application of this method will be further considered in the next lesson.

EXERCISES.

1. If a parallelogram be suspended at a point in one of its sides, what point will be vertically under the point of suspension?
2. A circular tower, the diameter of which is 20 ft., is being built, and for every foot it rises it inclines 1 in. from the vertical; what is the greatest height it can reach without falling?
3. A circular table weighs 20 lbs. and rests on four legs in its circumference forming a square; find the least pressure that must be applied at its edge to overturn it.

4. Two weights of 17 lbs. and 13 lbs. are connected together by a weightless rod; at what point must the rod be supported to rest in any position whatever?

5. A solid cube rests on a rough plane; through what angle may the plane be inclined before the cube overturns?

6. An equilateral triangle stands vertically on a rough plane; find the ratio of the height to the base of the plane when the triangle is on the point of overturning.

7. A triangular board weighing 30 lbs. is carried by 3 men, each standing at one of the corners; what weight does each bear?

8. Four weights of 3 ozs., 4 ozs., 5 ozs., and 6 ozs. are placed at the corners of a square; find the position of their centre of gravity.

9. To one corner of a heavy square an equal weight is attached; where must it be suspended by a single string to rest horizontal?

10. Find the centre of gravity of a triangular lamina, to two corners of which weights, respectively equal to half the weight of the triangle, are attached.

11. A ladder 20 ft. long weighs 60 lbs.; its centre of gravity is 8 ft. from the thicker end; it is carried by two men, one of whom supports the heavier end on his shoulder; where must the other stand that the weight may be equally divided?

12. Weights of 1 oz., 2 ozs., 3 ozs. are placed at the corners of an equilateral triangle; find their centre of gravity.

13. An isosceles triangle rests on a square, and the height of the triangle is equal to a side of the square; find the centre of gravity of figure thus formed.

14. Find the centre of gravity of a figure in the shape of a kite, formed by two isosceles triangles put base to base. The extreme width of the kite is 4 ft., and the heights of the isosceles triangles are 2 ft. and 4 ft. respectively.

15. Two particles weighing 4 ozs. and 10 ozs. are joined together by a uniform rod weighing 3 ozs.; find at what point the rod must be supported to rest in any position whatever.

16. Find the inclination of a rough plane on which half a regular hexagon can just rest in a vertical position without overturning, with the smaller of its parallel sides in contact with the plane.

XXVI.—*Methods of finding the Centre of Gravity of bodies joined together and of parts of bodies.*

§ 189. **Given the Centre of Gravity of each of two bodies, to find the Centre of Gravity of both together.**—The centre of gravity of both bodies can be found by joining the centres of gravity of the two, and dividing the line thus drawn into two parts in the inverse ratio of the weights of the bodies. The process is the same as in finding the centre of gravity of two heavy particles.

§ 190. **Example.**—Two cylinders with circular bases have a common axis; the radius of the one is 4 inches and the altitude 6 inches; the radius of the other is 3 inches and the altitude 8 inches; find the centre of gravity of both.

Since the volume of a cylinder is equal to $\pi r^2 h$, where r is radius of the base, and h the height, the volume of the one cylinder is $\pi \times 16 \times 6$ and of the other $\pi \times 9 \times 8$.

The distance $g_1 g_2 = 7$ inches. If G be the centre of gravity required

$$\frac{g_1 G}{g_1 g_2 - g_1 G} = \frac{72}{96} \therefore g_1 G = \frac{9}{21} g_1 g_2$$

$$\therefore g_1 G = 3 \text{ inches};$$

or their centre of gravity is at the point in the axis where they touch.

FIG. 114.

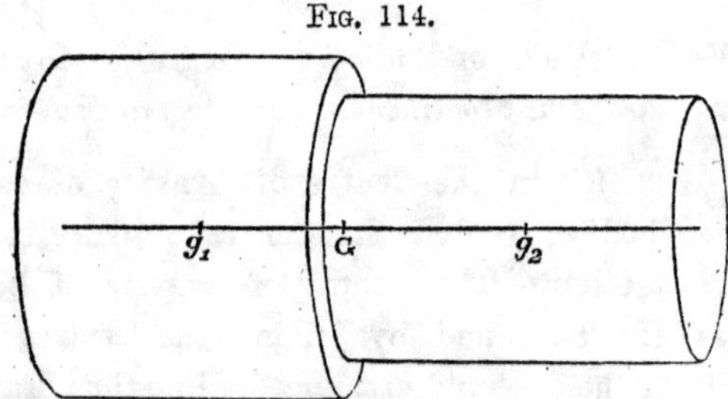

The centre of gravity of several bodies joined together can be found in the same when their several centres of gravity lie on the same straight line; otherwise the problem generally presents numerical difficulties.

§ 191. **Example.**—To find the centre of gravity of three uniform rods joined together so as to form the letter F, the lengths of the rods being as 1 : 3 : 6, and their thickness neglected.

Let the length of the vertical rod be 6 inches, the horizontal rod 3 inches, and the small central rod 1 inch.

Take $g_1 g_2$ and divide it in G', in the ratio of 1 : 2. Then the centre of gravity of these two rods is at G'. Draw a vertical through G'. This bisects the central rod and passes, therefore, through its centre

FIG. 115.

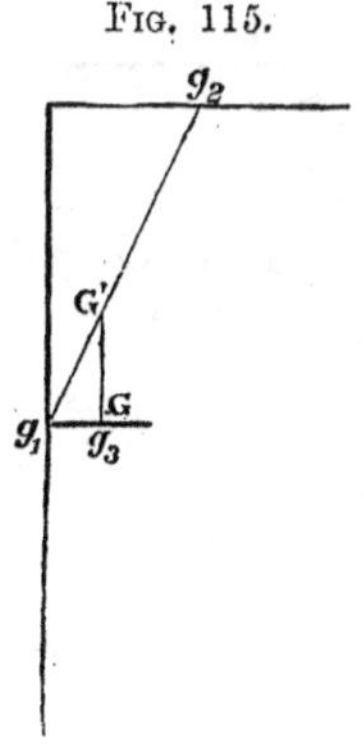

of gravity. Mark off $\frac{1}{10}$ of $g_3 G'$ from g_3, and the point G thus determined is the centre of gravity required.

§ 192. **Given the Centre of Gravity of a whole figure and of a part of it cut off, to find the Centre of Gravity of the remainder.**—Let W be the weight of the whole figure and C its centre of gravity. Let w be the weight of a part, and g its centre of gravity. Then $W-w$ is the weight of the remainder.

Join Cg and let the centre of gravity of the remainder act at some point G.

Fig. 116.

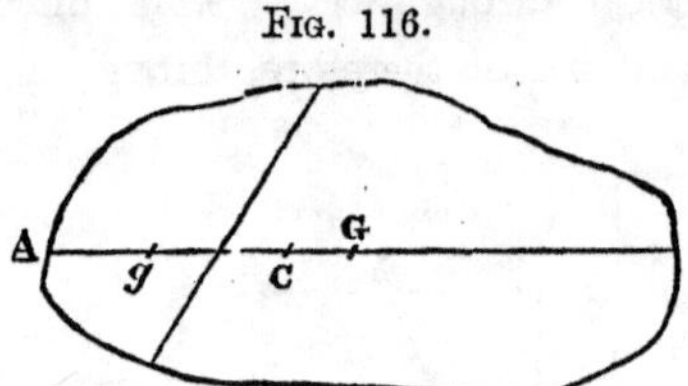

Produce Cg to some point A. Then, by the principle of moments,

$$W \times AC = w \times Ag + (W - w) \times AG$$

$$\therefore\ AG = \frac{W \times AC - w \times Ag}{W - w}.$$

§ 193. **Example.**—To find the centre of gravity of the remainder of a square, out of which, one of

Fig. 117.

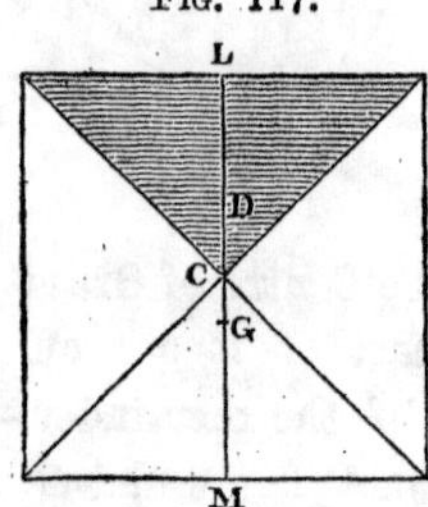

the triangles formed by the diagonals has been taken.

The figure that remains is symmetrical about

the line LM. The centre of gravity is therefore in that line.

Let the side of the square be a. Then, the area of the whole square is a^2 and its centre of gravity is at C; the area of the triangle is $\frac{a^2}{4}$ and its centre of gravity is at D; and the area of the remainder is $\frac{3a^2}{4}$ and let its centre of gravity be at G.

Taking moments about M, we have

$$a^2 \times MC = \frac{a^2}{4} \times MD + \frac{3\,a^2}{4} MG$$

$$\therefore \frac{a}{2} = \frac{1}{4}\left(\frac{a}{2} + \frac{a}{3}\right) + \frac{3}{4} MG$$

$$\therefore MG = \tfrac{7}{18}\, a,$$

or G is situated $\frac{1}{9}$ of a from C.

The centre of gravity of this figure might have been easily determined by finding the centres of gravity of each of the triangles of which it is composed.

§ 194. **General method of finding the Centre of Gravity of a number of particles or bodies.**—Let w_1, w_2, w_3 . . . be the weights of several particles or bodies, x_1, x_2, x_3 . . . the distances of their centres of gravity from a fixed line Oy; and y_1, y_2, y_3 . . . the corresponding distances from a line Ox at right angles to Oy; then if X and Y be the respective distances of their centre of gravity from these lines

$$X = \frac{w_1 x_1 + w_2 x_2 + w_3 x_3 + \ldots}{w_1 + w_2 + w_3 + \ldots}$$

and $$Y = \frac{w_1 y_1 + w_2 y_2 + w_3 y_3 + \ldots}{w_1 + w_2 + w_3 + \ldots}$$

Let A and B be the centres of gravity of two particles or bodies whose weights are w_1 and w_2.

FIG. 118.

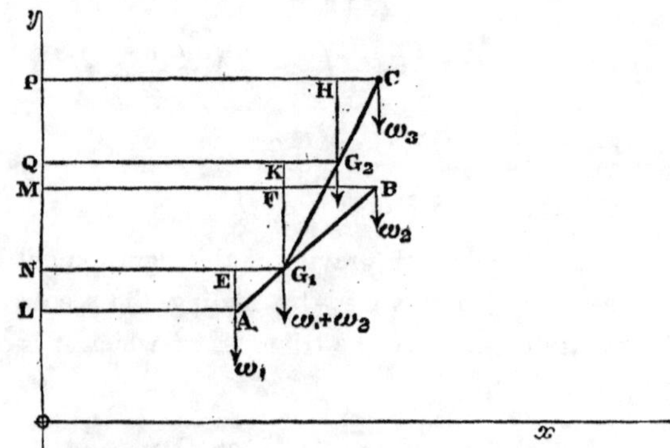

Let the resultant of w_1 and w_2 act at G_1. Draw the lines AL, G_1N, BM parallel to Ox, and AE, G_1F parallel to Oy.

Then $$w_1 \times AG_1 = w_2 \times BG_1$$

$$\therefore w_1 \times AE = w_2 \times G_1F$$

or $$w_1 \times (ON - OL) = w_2 \times (OM - ON)$$

$$\therefore (w_1 + w_2)\, O N = w_1 \times O L + w_2 \times O M$$

or $$O N = \frac{w_1 \times O L + w_2 \times O M}{w_1 + w_2}.$$

Let C be the centre of gravity of another body, the weight of which is w_3. Join $C G_1$.

Let G_2 be the centre of gravity of $(w_1 + w_2)$ at G_1 and of w_3 at C. Draw $G_2 Q$, $C P$ parallel to $O x$. Produce $G_1 F$ to K, and draw $G_2 H$ parallel to $O y$.

Then $$(w_1 + w_2) \times G_1 G_2 = w_3 \times G_2 C$$

$$\therefore (w_1 + w_2) \times G_1 K = w_3 \times G_2 H$$

$$\therefore (w_1 + w_2)(O Q - O N) = w_3 \times (O P - O Q)$$

$$\therefore (w_1 + w_2 + w_3)\, O Q = (w_1 + w_2)\, O N + w_3 \times O P$$

$$= w_1 \times O L + w_2 \times O M + w_3 \times O P$$

$$\therefore O Q = \frac{w_1 \times O L + w_2 \times O M + w_3 \times O P}{w_1 + w_2 + w_3}$$

In the same way it may be proved for any number of points that $X = \dfrac{w_1 x_1 + w_2 x_2 + w_3 x_3 + \ldots}{w_1 + w_2 + w_3 + \ldots}$ where X is the distance of their centre of gravity from $O y$. If we draw lines from A, B, and C parallel to $O y$ and measure along $O x$, the same kind of reasoning will prove that

$$Y = \frac{w_1 y_1 + w_2 y_2 + w_3 y_3 + \ldots}{w_1 + w_2 + w_3 + \ldots}$$

Where Y is the distance of the centre of gravity of the whole system from the line $O x$.

By the aid of this proposition, which is the most

general we have proved, the centre of gravity of several particles that do not lie in the same straight line can conveniently be found.

§ 195. If the particles are not in the same plane, and z_1, z_2, z_3, be their respective distances from a fixed plane, it is equally true that

$$Z = \frac{w_1 z_1 + w_2 z_2 + w_3 z_3 + \dots}{z_1 + z_2 + z_3 + \dots}$$

where Z is the distance of their centre of gravity from the plane.

§ 196. **Example.**—From the point D (fig. 119) which bisects $A\,C$, one of the sides of the equilateral triangle $A\,B\,C$, a straight line $D\,E$ is drawn perpendicular to the base cutting off the triangle $D\,E\,C$. Required the centre of gravity of the figure $A\,D\,E\,B$.

Take $B\,C$ as one fixed line, and $B\,C'$ at right angles to it. Find g_1, the centre of gravity of the whole triangle, and g_2 of the part. Then $B\,F = x_1$ and $B\,K = x_2$ and $F\,g_1 = y_1$, and $K\,g_2 = y_2$.

The area of the whole triangle equals $w_1 = \frac{a\,h}{2}$ where $B\,C = a$ and $A\,F = h$; the area of the small triangle equals $w_2 = \frac{w_1}{8} = \frac{a\,h}{16}$; the area of the remaining figure equals $w_1 - w_2 = \frac{7}{16}\,a\,h$.

Then, if G be the centre of gravity required, and $B\,M = X$, and $G\,M = Y$, we have,

$$X = \frac{\frac{a\,h}{2} \times BF - \frac{a\,h}{16} BK}{\frac{7}{16} a\,h}$$

$$= \frac{8\,BF - BK}{\;};$$

and $\qquad BF = \frac{a}{2};$ and $BK = \frac{5}{6} a.$

$$\therefore X = \frac{4\,a - \frac{5}{6} a}{7} = \frac{19}{42} a.$$

Fig. 119.

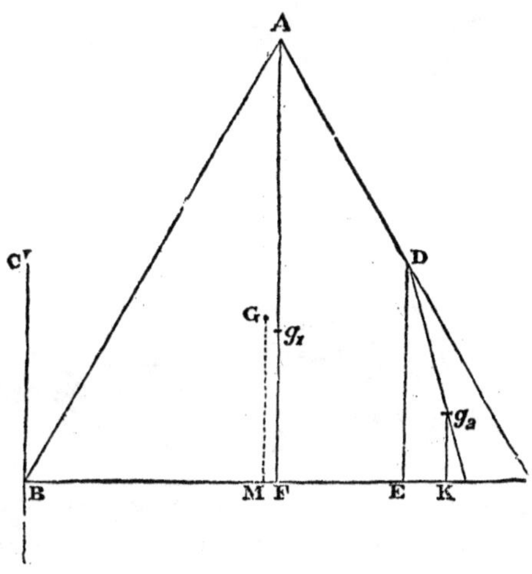

Similarly, $\quad GM = Y = \frac{8\,Fg_1 - Kg_2}{7};$

and $\qquad Fg_1 = \frac{1}{3} h$ and $Kg_2 = \frac{1}{6} h;$

$$\therefore Y = \frac{\frac{8}{3}h - \frac{1}{6}h}{7} = \frac{15}{42}h.$$

Hence the centre of gravity is at a point G, where $BM = \frac{19}{42}a$ and $GM = \frac{15}{42}h$.

§ 197. **To find the Centre of Gravity of the Surface of a Cone.**—If two points a and b be taken in the circumference of the base of the cone, and each of these points be joined with C, the vertex,

Fig. 120.

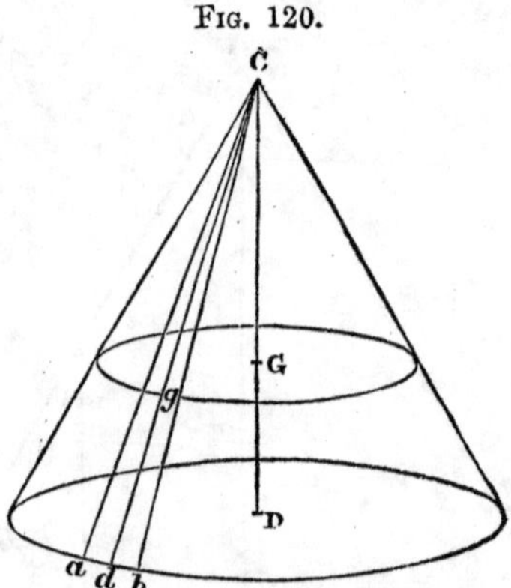

the figure thus formed may be regarded as a triangle if the points a and b be sufficiently near one another. The centre of gravity of the surface of this triangle is at a point g, one-third up the median line Cd. Now, the circumference of the base may be regarded as a polygon with an infinite number of infinitely small sides; and the surface of the cone may therefore be supposed to consist of an infinite number of such triangles, as $C\,a\,b$. If a circle be drawn through g parallel to the base, the centre of gravity of each of these triangles will lie on the circumference of this circle, and therefore the centre of this circle will be the point at which the weight of the sum of these triangles acts, *i.e.*, the centre of gravity of the surface of the cone. But the centre of this circle is at G, where $D\,G$ is equal to one-third of $D\,C$; therefore the centre of gravity of the surface of a cone is at a point in the axis at a distance of *one-third* of the axis from the centre of the base.

§ 198. If r be the radius of the base, and h the height of the cone, the area of the surface can be shown to be equal to $\pi\, r \sqrt{h^2 + r^2}$, or half the circumference of the base into the slant side.

§ 199. **Centre of Gravity of a Pyramid.**—Since a pyramid may be supposed to consist of a number of parallel discs or flat prisms of the same shape as the base, lying one on the other, the centre of gravity of all these discs, and therefore of the pyramid, must be situated in the line joining the centre of gravity of the base with the apex of the pyramid, and it can easily be shown to be at a distance of *one-fourth* of this line from the base.

Thus, if C be the centre of gravity of the base of a pyramid, and A be its apex, and if $A\ C$ be joined,

Fig. 121.

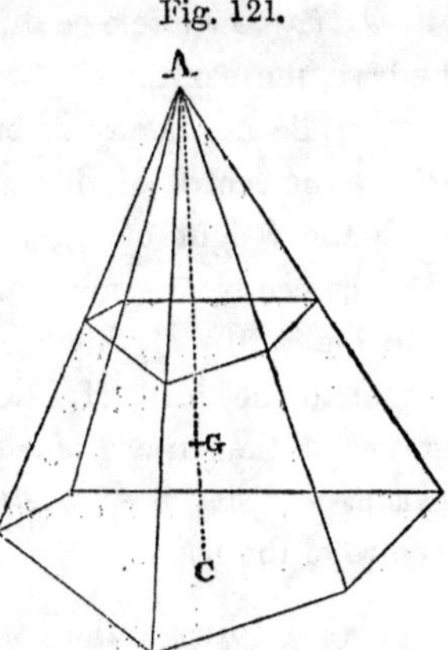

and $C\ G$ be marked off equal to one-fourth of $C\ A$, then G is the centre of gravity of the pyramid.

§ 200. **Centre of Gravity of a Cone.**—If the number of sides of a pyramid be increased without limit, the pyramid becomes a cone. The centre of gravity of a cone is found, therefore, in the same way as that of a pyramid, by joining the apex of the cone with the centre of gravity of its base, and measuring one quarter of this line from the base.

EXERCISES.

1. A heavy wire is bent in the form of a right angle, and one arm is twice the length of the other. At the angular point a weight equal to half the weight of the whole wire is fixed; find the centre of gravity of the system.
2. A uniform rod weighs 8 ozs., and to one end is attached a weight of 12 ozs., and to the other a weight of 6 ozs.; the rod is 20 ins. long; where must it be supported to rest in a horizontal position?
3. A cylinder the diameter of which is 10 ft., and height 60 ft., rests on another cylinder, the diameter of which is 18 ft., and height 6 ft.; and their axes coincide; find their common centre of gravity.
4. An equilateral triangle rests on a square, and the side of the triangle is equal to the side of the square; find the centre of gravity of the figure thus formed.
5. Into a hollow cylindrical vessel 11 ins. high, and weighing 10 ozs., the centre of gravity of which is 5 ins. from the base, a uniform solid cylinder 6 ins. long, and weighing 20 ozs., is just fitted; find their common centre of gravity.
6. From a uniform circular disc another disc, having for its diameter the radius of the first circle, is cut away; find the centre of gravity of the remainder
7. From a square the side of which is 6 ins. a corner square is cut away, the side of which is 2 ins.; find the centre of gravity of the remainder.

EXAMINATION.

1. Define the centre of gravity of a body, and give an experimental way of finding the centre of gravity of a uniform lamina.

2. A heavy wire is bent at its middle point, so as to contain an angle of 60°; it is suspended from one of its ends; find its position in equilibrium.
3. Explain how it is that a long rod is more easily balanced on its end than a short one.
4. $ABCD$ is a square; CB is produced to E, and on EB is described a square. If $BE = \frac{1}{3} AB$, find the centre of gravity of the figure thus formed.
5. Find the height of a cylinder which can just rest on an inclined plane the angle of which is 60°; the diameter of the cylinder being 6 ins.
6. Define stable, unstable, and neutral equilibrium. Can a body have more than one position of stable equilibrium?
7. How is the potential energy of a body connected with its state of equilibrium?
8. From a rectangle 8 ins. by 5 ins. a corner rectangle is cut out 2 ins. by 3 ins. (2 ins. from the side of 8 ins.); find the centre of gravity of the remainder.
9. A rectangular board 6 ins. by 8 ins., and weighing 2 ozs., is hung up from one of its angular points; to the extremity of the adjacent short side a weight of 8 ozs. is attached; find the position in which the board hangs in equilibrium.
10. A cylindrical vessel weighing 4 lbs., and the internal depth of which is 6 ins., will just hold 2 lbs. of water. If the centre of gravity of the vessel when empty is 3·39 ins. from the top, determine the position of the centre of gravity of the vessel and its contents when full of water.—*Matriculation, Univ. Lond.*, Jan. 1869.
11. Two uniform cylinders of the same material, one of them 8 ins. long and 2 ins. in diameter, the other 6 ins. long and 3 ins. in diameter, are joined together, end to end, so that their axes are in the same straight line. Find the centre of gravity of the combination.—*Ib.*, June 1869.
12. Two heavy particles, weighing respectively 3 and 5 ozs.,

are attached to the ends of a straight rod 8 ins. long, weighing 2 ozs; find the centre of gravity of the system.—*Matriculation, Univ. Lond.*, Jan. 1870.

13. What condition must be fulfilled in order that a body may be in equilibrium upon a hard smooth plane surface?—*Ib.*, June 1870.

14. Four heavy particles of the relative weights 2, 3, 4, 5 are placed at the corners A, B, C, D respectively of a horizontal square board; find the common centre of gravity of the four particles.

Show whether the centre of gravity would change if the board were inclined.—*Ib.*, Jan. 1871.

15. Give examples of bodies in stable, unstable, and neutral equilibrium. If a body be in stable equilibrium how is its centre of gravity affected by a small displacement of the body?—*Ib.*, June 1873.

16. A trapezium, having two parallel sides, which are 4 and 12 ft. long, and the other sides each equal to 5 ft., is placed with its plane vertical, and with its shortest side on an inclined plane; find the relation between the height and base of the plane when the trapezium is on the point of falling over.—*Preliminary Scientific 1st M. B.*, 1870.

17. Find the centre of gravity of three equal rods, AB, AC, and AD, in the same plane, and diverging from the point A, each of the angles BAC and CAD being one-third of a right angle.—*Ib.*, 1871

18. Weights of 2, 3, 2, 6, 9, 6 kilogrammes are placed at the angular points of a regular hexagon taken in order; determine the position of their centre of gravity.—*Ib.*, 1874.

APPENDIX.

A.

EXAMINATION PAPERS SET AT VARIOUS INSTITUTIONS.

1. *Matriculation Examination, January* 1875. (*Friday, January* 15.)

Natural Philosophy.

1. State and exhibit the mechanical conditions of equilibrium, when a beam *A B*, of uniform thickness and density, rests with one extremity against a smooth vertical wall, and the other against a smooth hemispherical bowl. [§§ 167–8.]
2. In a system of pulleys in which a separate string passes round each pulley, it is found that when the power and the weight are in equilibrium and the power is caused to descend through 8 ft., the weight rises through 1 ft. Can the mechanical advantage of this arrangement ever be as much as 8? Give reasons for your answer. [§§ 83, 97.]
3. Equal weights (each 1 oz.) are placed at the angular points of a heavy triangular lamina and also at

the middle points of the sides. Find the position of the centre of gravity of the plate and weights.

4. A river one mile broad is running downwards at the rate of four miles an hour, and a steamer moving at the rate of eight miles an hour wishes to go straight across. How long will the steamer take to perform the journey, and in what direction must she be steered? [§ 24.—8·66 mins.]

5. A rifle-bullet is shot vertically downwards from a balloon at the rate of 400 ft. per second. How many feet will it pass through in two seconds, and what will be its velocity at the end of that time, neglecting the resistance of the air, and estimating the acceleration due to gravity at 32? [864; 464 ft.]

6. It is stated that, because the earth rotates on its axis, a lump of lead weighs less in Ceylon than in Spitzbergen. Explain the statement. [§§ 29, 46.]

---

2. *Science and Art Department of the Committee of Council on Education, South Kensington.*

Theoretical Mechanics.

1. Where is the centre of gravity of a triangle situated?

2. A body is, in shape, a sphere, but loaded in such a manner that its centre of gravity is not at its geometrical centre; when it is placed on a horizontal plane, what are its positions of stable and unstable equilibrium?

3. Three forces of 2, 10, and 12 units act along parallel lines on a rigid body; show how they may be adjusted so as to be in equilibrium.

4. What is meant by the sensibility of a balance?
   Other conditions being the same, why is the sensibility less the greater the weight of the beam?
5. Two forces of 10 units each act on a body along parallel lines, and in opposite directions; why would it be impossible to balance these forces by any one force?
6. *A CB* is a bent lever with its fulcrum at *C*, the angle *A CB* is a right angle, the arms *A C* and *B C* are 10′ and 7′ long, and *A C* is in a vertical position; a horizontal force of 21 units, acting at *A*, is balanced by a vertical force (*P*) acting at *B*; find the magnitude of *P*, and the pressure on the fulcrum, and show in a diagram the line along which the latter force acts.
7. State exactly what is meant when the velocity of a body is said to be uniformly accelerated. The acceleration of a body's velocity is denoted by 5, the units being feet and seconds; what fact is expressed by this number 5?
8. A body moves from a state of rest, and at the end of 8 secs. has a velocity of 30′ a second; its velocity is known to have been uniformly accelerated; how far did it go in the 8 secs.? Supposing the motion to continue under the same conditions, how far will it go in the next 8 secs.?
9. The mass of a body is 10 lbs.; at a given instant it is moving at the rate of 40′ a second; from this instant a constant force is made to act on it in a direction opposite to that of the motion, which brings it to rest after it has described 18′; what was the magnitude of that force? What unit of force have you used in stating the answer?

10. The force of gravity on a given mass is sensibly different at different places on the earth's surface; state briefly how this can be shown to be the case.

21. Investigate a formula for the position of the centre of gravity of a trapezoid, having given the lengths of the two parallel sides, and that of the line joining their middle points.

22. In the case of two forces which act along intersecting lines, show that the sum of the moments of the forces with respect to any point in the plane of the lines equals the moment of their resultant with respect to the same point. [It will be enough to consider one case.]

What assumption must be made respecting the signs of the moments, in order that the above statement may include all cases?

23. A heavy point is suspended by a thread from a fixed point, and rests against a smooth inclined plane; the inclination of the thread to the vertical and that of the plane to the horizon is 30°; determine the pressure on the plane and the tension of the thread in terms of the weight of the point (W).

24. A body rests on a rough horizontal plane; determine the direction of the smallest force that would make the body slide. Under what circumstances would a smaller force overthrow the body?

26. An inclined plane has a base 120′ long and 50′ high; the co-efficient of friction between it and a body weighing 56 lbs. placed on it is 0·5; how many units of work are required to draw the body up the plane, and how many to draw it down the plane?

27. Two bodies, whose masses are 2 cwt. and 96 lbs. respectively, move from rest under the action of constant forces; the former is found to describe 8′ in the first three seconds of its motion, the latter 5′ in the first two seconds of its motion; find the ratio of the former force to the latter.

28. A body falling from rest and moving freely in vacuo describes 336′ in one second; for how long had it been moving before the beginning of that second? ($g=32$).

---

3. *College of Preceptors. Pupils' Examination, Christmas,* 1874.

Mechanics.

1. State the proposition termed the Triangle of Forces.

   A triangle *A B C*, formed of three iron rods, has the points *A* and *B* fixed to two points in a vertical post, and a weight of 1 ton is suspended from *C*: having given $AB=4$ ft., $AC=20$ ft., and $BC=18$ ft., find the tension in *B C*.

2. A rod 4 ft. long has a weight of 2 lbs. at one end, 3 lbs. at 15 in. from the end, 4 lbs. at 27 in., 5 lbs. at 40 in., and 8 lbs. at the other end; find at what point the rod must be supported to remain horizontal.

3. Mention the principal properties of the Centre ot Gravity of a body.

   A solid whose ends are squares 21 in. by 21 in., and whose faces are parallelograms, the length of the shorter side of each being 21 in., and that of the longer 35 in., is just on the point of falling over when placed on a level table. Find the height of the solid.

4. What power must a man weighing 150 lbs. exert to lift himself by a pair of pulley-blocks, each containing two wheels?
5. Find the relation between the power and the weight when a load is sustained by a screw turned by a lever.
6. A body is projected vertically upwards with a velocity of 60 ft. per sec. (i) How long will it be rising 11 ft.? (ii.) What will be its greatest elevation? (iii.) How long will it be before it returns to the starting-point?
7. State Newton's Second Law of Motion.

   A weight of 16 lbs. is attached by a string to the car of a balloon which is ascending with an acceleration of 10 ft. per sec.; find the tension of the string.

---

4. *University of Oxford. Local Examination, Thursday, June* 4, 1874. *For Junior Candidates.*

Mechanics and Mechanism.

1. How is the resultant of two given forces found geometrically, (1) when they act at the same point and include a given angle; (2) when they are parallel and act in opposite directions?

   The resultant of two forces inclined at an angle of 135° is 2, and one of the forces $2\sqrt{2}$; find the other.
2. Enunciate the principle of the Triangle of Forces.

   A weight of 10 lbs. is suspended by two strings

attached at their upper ends to two points in the same horizontal line; their lengths being 39 and 52 in. respectively, and the angle between them being 90°; find the tension in each.

3. Define Centre of Gravity.

Find that of a plane figure composed of a square with an equilateral triangle described on one of its sides.

4. Determine the inclination of a smooth inclined plane when a force of 5 lbs. along it will just sustain a weight of 10 lbs. upon it.

What would the coefficient of friction be, supposing the plane just rough enough to hinder the weight slipping down when left to itself?

5. Describe Attwood's machine.

It is found that when the weights are 119 and 121 grs. respectively the distance passed over in the third sec. is 8 in.; determine the value of g.

6. A stone is thrown vertically upwards with a velocity of 48 feet per second; find how high it will ascend, with what velocity it will return to the earth, and the whole time of motion.

For Senior Candidates.

1. Define the terms—Moment of a Force, Momentum, Angular Velocity.

Assuming the parallelogram of forces, find the conditions of equilibrium of any number of forces acting at one point and in the same plane.

2. Five forces acting at one of the angular points of an equilateral and equiangular hexagon are represented in direction and magnitude by the straight

lines which join that point to the other angles of the figure; find the resultant.

3. If parallel forces of given magnitude are applied at fixed points, show that their resultant passes through a point the position of which is independent of the direction of the forces.

4. Find the centre of gravity of the remaining portion of a square, when one corner has been cut off along a line joining the middle points of two adjacent sides.

5. If a right circular cone will just rest on its side upon a table with its vertex projecting over the edge of the table to the distance of $\frac{1}{3}$ of the slant side, find the ratio of the altitude to the radius of the base.

6. A weight of 1 cwt. is supported on an inclined plane, whose inclination to the horizontal plane is 45°, by a string which passes over a pulley at the top of the plane and has a weight attached to it. Find the least possible value of the supporting weight when the coefficient of friction is $\frac{1}{2}$, and its value when the plane is smooth.

7. Explain the principle of 'Virtual Velocities.'

   Find the relation between the power and weight in the cases of the lever, screw, and windlass.

8. How is it shown that the velocity of a body, moving from rest in vacuo under the action of a constant accelerating force, is proportional to the time it has been in motion, and that the space passed through is proportional to the square of the time?

9. A particle is let fall from a given height, and at the same moment another particle is projected vertically upwards; if the two particles are at the

same height just as the second comes to rest, find how long one will reach the ground before the other.

---

5. *University of Cambridge. Local Examinations, December* 1873

MECHANICS.

1. Define force, and show that forces may be properly represented by straight lines.
2. Enunciate the theorem known as the 'Triangle of Forces,' and deduce it from that known as the 'Parallelogram of Forces.'

   A rod is supported by two strings attached to its extremities. One of the strings is fastened to a small ring, and the other passes through the ring and is gradually lengthened; show that the centre of gravity of the rod will describe a straight line.
3. Define the term 'Centre of Gravity.'

   The centre of gravity of a body and of a portion of it being given, find the centre of gravity of the remainder.

   A circular plate has a circular piece removed from it, the diameter of the piece being half that of the plate; find the position of the centre of gravity of the remainder, the edge of the piece removed passing through the centre of the plate.
4. A rod of uniform thickness has one-half of its length composed of one metal, and the other half of a different metal. The rod will balance about a point distant one-third of its whole length from

one extremity. Compare the weights of equal volumes of the two metals.

5. Describe the common (Roman) steelyard and explain the mode of graduating it.
6. Find the ratio of the power to the weight in that system of pulleys in which a separate string passes round each pulley, the weights of the pulleys being neglected.

   If the power be equal to the weight, find the weight of a pulley in the above system, assuming that all the pulleys are of the same weight.

---

6. *Royal College of Surgeons of England. Preliminary General Examination, Midsummer* 1874.

Mechanics.

1. Explain clearly the mode of *representing* pressures by lines, and how this mode is used in finding the *resultant* (i.) of two pressures acting at a point, (ii.) of three pressures acting at a point.
2. A peg driven into the centre point of a square wall of a room is pulled at with two strings, the pull on one being represented by the vertical from the peg to the ceiling, that on the other being represented by the slant line from the peg to the right hand lower corner of the wall. Find (i.) the line representing the force which, if applied when the other two were removed, would produce just the same stress on the peg as they by their joint action do; and (ii.) the line representing the force which, if applied with the other two, would take all stress off the peg.

3. It is said that the 'six simple machines are the Lever, the Wheel-and-Axle, the Pulley, the Inclined Plane, the Wedge, and the Screw.' Point out that some of these machines are merely modified forms of others. Why are they sometimes termed Mechanical Powers?
4. Sketch any system of two moveable pulleys, and point out how, by using such a system, mechanical advantage would be derived.
5. Distinguish the three different kinds of Levers, and give an example of each, describing it so as to show clearly that it is an example.
6. State the conditions for the equilibrium of pressures acting on a lever.

   If a walking-stick be supported near one end by passing that end over the thumb and under the fourth finger, it is noticeable that the pressure perpendicular to the stick which must be exerted by the fourth finger is greater when the stick is held horizontally than when in any slanting position. Explain this fact clearly on mechanical principles, with the help of a diagram.
7. Define the term Centre of Gravity. How is the idea of the centre of gravity used in making calculations where the weights of bodies are to be taken into account?

   A scaffold-pole 60 ft. long balances on a log put under it 35 ft. from one end, and it also balances on the log put under its centre when a boy weighing 90 lbs. is sitting on it at one end and a man weighing 160 lbs. at the other; find in round numbers the weight of the pole.
8. Explain clearly on mechanical principles why a loaded wagon going obliquely across a declivity

may turn over, while if the same load could be compressed, the wagon might safely pass the same spot in the same course.

9. A machine is started vertically into the air with a velocity of 60 ft. per minute, having a flying apparatus attached to urge it vertically upwards, and its motion is observed to be uniformly retarded by 10 ft. per minute. How high will the machine rise?

---

7. *University of Cambridge. Previous Examination, December* 12, 1874.

Mechanics.

1. Define the resultant of any number of forces.

   If a system of forces be in equilibrium, prove that each of these forces is equal to the resultant of all the rest, and acts in a direction directly opposite to the direction of that resultant.

2. Three forces in equilibrium act at the same point; prove that each force is proportional to the sine of the angle between the other two forces.

   P and Q are two forces acting at a point, and R is their resultant; S is the resultant of R and P. Prove that, if the forces P and Q be inclined to each other at a right angle, and if $Q=3P$ then $S=P\sqrt{13}$.

3. Define the *centre of gravity* of a body. Prove that if the centre of gravity be found for one position of a body, it will be the centre of gravity when the body is turned into any other position.

4. If a body be placed on a horizontal plane, prove that the body will stand or fall according as the ver-

tical through its centre of gravity falls within or without the base on which it rests.

5. What are the requisites for a good balance?

A tradesman's balance has arms whose lengths are 11 in. and 12 in. respectively, and it rests horizontally, when the scales are empty. If he sell to each of two customers a pound of tea at 2*s.* 9*d.* per pound, putting his weights into different scales for each transaction, find whether he gains or loses owing to the incorrectness of his balance, and how much.

6. Find the ratio of the power to the weight in a system of pulleys, each pulley being suspended by a separate vertical string, which has one end fixed to a beam above. [The weight of the pulleys may be neglected.]

7. Describe the Wheel-and-axle.

If the string from the axle pass round a moveable pulley and have its end fixed to a beam above, the two parts of the string being parallel, and if the weight be attached to the moveable pulley, find the ratio of the power to the weight.

8. A straight uniform rod 12 in. long is suspended by two vertical strings attached to the ends of the rod; at a distance of 2 in. from one end a weight of 7 pounds is attached, and a weight of 1 pound at the same distance from the other end. Find the greatest weight the rod can have without breaking the strings. The strings break when subjected to a tension of more than 7 pounds.

Matriculation Examination, June 1875.

1. Two parallel forces, P and Q, act at two points in a straight line, 6 inches apart, in opposite directions. Their resultant is a force of 1 lb. acting at a point in the line 4 feet from the larger of the forces P and Q. Determine the values of P and Q. [8 lbs.; 9 lbs.]
2. A short circular cylinder of wood has a hemispherical end. When placed with its curved end on a smooth table it rests in any position in which it is placed. Determine the position of its centre of gravity. [At the centre of the sphere.]
3. A uniform circular arc rests in a vertical plane on two smooth pegs, B and C. Find by aid of a sketch the mechanical condition of equilibrium. [§§ 59, 167.]

B C

4. What is meant by saying with reference to gravity $g=32$? What would be the value of g if your units of space and time were miles and minutes? [§ 29.—$21\frac{9}{11}$.]
5. A stone is let fall from the top of a railway-carriage which is travelling at the rate of 30 miles an hour. Find what horizontal distance and what vertical distance the stone will have passed through in one-tenth of a second. [4·4 ft.; 0·16 ft.]
6. At the earth's equator the hot air ascends, and is replaced by cold air which blows in along the ground from the poles. That which comes from our hemisphere blows from the north-east instead of from the north. Explain this. [§ 22.]

First B. Sc. and Preliminary M.B. Pass Examinations, 1875.

MECHANICAL PHILOSOPHY.

1. Define the expression 'Centre of Parallel Forces.' Obtain the position of the centre of four parallel forces, P, $2P$, $3P$, $4P$, acting at the angular points of a square taken in order. [§ 141; § 143. Ex. (3).]
2. A perfectly flexible heavy chain of uniform density and section is made to hang as in the figure. Show by means of a sketch the arrangement of the forces which keep any portion of the chain, AB, in equilibrium. [§ 167.]

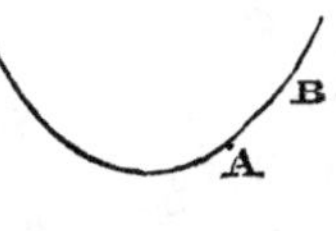

3. In a large hotel there is a lift-chamber, from the roof of which a weight hangs by means of an india-rubber string. Suddenly the support of the lift-chamber gives way, and the chamber, with all that it contains, falls freely down under the action of gravity. What will now first happen to the weight and its india-rubber support? [See § 52. Ex. (9).]
4. A mass of 200 grammes is acted on by a force equal to the weight of 10 grammes for 20 seconds. What distance will the mass have passed through, and what velocity will it have acquired?

   [The acceleration due to gravity is 980 (centimetres, seconds)]. [s = 9,800 c.m.; v = 980 cent. secs.]
5. Prove that the time of descent of a heavy particle down all chords from the highest point of a vertical circle is the same. [§ 37.]

B.

ANSWERS TO EXERCISES AND EXAMINATION QUESTIONS.

I. Exercises. Pp. 19–20.

(1) 40 secs. (2) 2·4 ft. (3) 120 ft. per sec. (4) 20 secs. (5) 257·6 ft. (6) $f = 20$; $v = 200$. (7) 200 ft. per sec. (8) 195 ft. (9) 40 ft. per sec. (11) 30 ft. per sec. (12) 60 ft. per sec.

II. Exercises. P. 30.

(1) No. (2) $10\sqrt{5}$ ft. (3) $16\sqrt{2}$ ft. per sec. (4) 595 yards. (5) $117\cdot\dot{3}$ secs.; 1,147·2 yards. (6) Straight line midway; vel. = 10.

Examination (Chapter I.) Pp. 37–9.

(3) $109\frac{1}{11}$ ft. from A. (4) 20 ins. per sec. (5) $2\frac{3}{11}$ miles from A. (7) 5 ft. per sec. (10) 50 ft. N. (12) $12\sqrt{13}$ ft. distant. (13) 33 ft. (15) 1,500 ft. (18) 90 ft. (19) $5\sqrt{3}$ ft. per sec. (20) $10\sqrt{2}$.

V. Exercises. Pp. 51–3.

(1) 100 ft. (2) 160 ft. per sec. (3) 176 ft.; 576 ft. (4) $\frac{25}{2}g$. (5) 5 secs. (6) 204 ft. (7) 178 ft.

(8) $16\sqrt{10}$ ft. per sec. (9) 100 ft. (10) 187·44 ft. (11) $67\frac{1}{3}$ ft. (12) 135 ft. (13) 336 ft. (14) 224 ft. (15) 10 ft. per sec. (16) 48,144,208 ft. (17) 88 ft. (18) $1\frac{7}{9}$ secs. (19) 96 ft. (20) 52·8 ft. (21) 96 ft.; 3 secs. (22) $8\sqrt{10}$ ft. per sec. (23) 38·4 ft. per sec. (24) 95·7 ft. nearly. (25) 824 yds. (26) 68·75 secs. (27) 312·5 ft.; 1.250 ft.

Examination (Chapter II.) Pp. 53–7.

(1) 25 ft. (5) $\frac{5g}{2}, \frac{9g}{2}, \frac{13g}{2}$. (6) $v = a + xg$. (8) $6\frac{1}{4}$ secs. (10) 106 ft. (11) 144·9 ft.; 16·1. (12) $2g\sqrt{5-2\sqrt{2}}$ ft. (13) 114 ft.; 144 ft. (14) 160 ft. (16) 140 ft. per sec. (17) 1,608 ft. per sec. (18) 74 ft. per sec. (19) 156 ft. (21) 64·28 ft. nearly. (24) $\frac{\sqrt{10}}{2}$ secs. (26) 402·5 ft.; 177·1 ft. (27) $174\frac{2}{9}$ ft. per sec.; 784 ft. If the balloon is still ascending when the stone is let fall v = 68·17 ft. per sec.; $s = 306{\cdot}76$ ft. (29) Three-quarters of the way up. (30) 144 ft. (31) 2 secs. (32) 225 ft.; $2\frac{1}{2}$ or 5 secs. (33) East. (34) 8·05 ft.; 120·75 ft. lower down. (36) 64 ft. (37) 36 ft.; 75 ft.; 3 secs.

VII. Exercises. Pp. 79–80.

(1) 32 ft. (2) 1 minute. (3) $\frac{1}{2}$ lb. (4) 3 ft. (5) 7·2 ozs. (6) 6 ozs. (7) $\frac{1}{2}$ oz. (8) 5 ozs. (9) 3·75 ozs. (10) $5\frac{13}{27}$ ozs. (11) $\frac{9}{64}$ ft.; 268 ft. (12) $4\frac{8}{9}$ mins. (13) 13·75 ozs. (14) 126 lbs.; 98 lbs. (15) $16\sqrt{2}$ ft. (16) 10 ft. per sec. (17) 1 minute.

VIII. Exercises. P. 84.

(1) 180 ft. (2) $266\frac{2}{3}$ ft. (3) 3 : 2.

EXAMINATION. (CHAPTER III.) Pp. 85-9.

(4) 32·1 : 31·4. (6) 1200g. (7) 6 ft. per sec. (10) 3 secs. (11) 3 secs. (12) 9 : 1. (13) 6 lbs. 9 ozs. (14) 10 mins. 25 secs. (15) 125 ft. (16) 2 secs. (17) $P = 14\frac{37}{52}$; $Q = 11\frac{23}{52}$. (19) 7 lbs. (20) 9 : 4. (22) 2·5 ft. (23) 5 : 1. (25) 2 : 3. (26) 40; $\frac{4}{9}$ mile. (27) $\frac{P}{W} = 11 : 1080$. (29) 2 tons 7 cwts. (30) 181·125 ft. (33) $4\frac{1}{4}$; $3\frac{3}{4}$. (34) 137·5 secs.; vel. = 10 miles an hour. (35) $\frac{16}{\sqrt{3}}$. (36) 48. (37) $\frac{g}{2}$; $\frac{\sqrt{3}}{2} \cdot g$. (38) $5\frac{164}{167}$ secs.

EXAMINATION (CHAPTER IV.) Pp. 99-102.

(3) 256 ft.; 176 ft. (5) $\frac{2\,P\,Q}{P + Q}$. (6) 3 : 2. (8) 50 lbs. (9) 12·5 ft. per sec. (10) 11 : 1. (12) $\frac{v}{2}, \frac{v}{3}, \frac{v}{4}$. (14) $3\frac{4}{7}$ ft. per sec. (15) 2 ft. (18) 12·8 ft. per sec. (19) 715 ft. per sec. nearly (21) 10, $6\frac{2}{3}$, 5, 4, &c.

XII. EXERCISES. P. 124.

(1) $\frac{3}{4}$. (2) 4 ozs. (3) 3·6 ozs. (4) 19,080ft.-lbs. (5) $\left(6 - \frac{12}{\sqrt{7}}\right) = 1{\cdot}46$ ozs. (6) 5,280,000 ft.-lbs. (7) 18,000 ft.-lbs. (8) 18,000 ft.-lbs. (9) 3,500 ft.-lbs. (10) 300 ft.-lbs.

EXAMINATION (CHAPTER V.) Pp. 125-6.

(5) $40\sqrt{10} = 126{\cdot}5$ ft. per sec. nearly. (6) 11,250 ft.-lbs.; $14\frac{221}{386}$ units of heat, Fahrenheit. (7) $\frac{W}{2g}(V^2 - u^2)$. (10) $1{,}444\frac{61}{106}$ thermal units Centigrade. (13) 1,080 ft.

XIV. Exercises. Pp. 131–8.

(1) 12 ins. from smaller weight. (2) 10 lbs.; 20 lbs. (3) 3 ft. from the heavier. (4) 21·6 ins.; 26 ozs. (5) 15 ozs. (6) 2 ins. from end where weight is. (8) 25 lbs. (9) $13\frac{1}{3}$ lbs. (10) 9·5 lbs.

Exercises. P. 142.

(1) 16 ins. (2) 8. (4) $3\frac{1}{3}$ ins. (5) 250 lbs. (6) 560 lbs. (7) The 8-oz. weight. (8) 240 lbs. (9) nt. (10) 120 lbs.

XV. Exercises. Pp. 152–3.

(1) 60 lbs. (2) 75 lbs. (3) 20 lbs. (4) 1 lb. (5) 27. (6) 211 ozs. or $213\frac{1}{4}$ ozs., as the 2 oz. pulley is highest or lowest. (7) 14 lbs. (8) 122 lbs. (9) $3\frac{3}{4}$ lbs. (10) $6{\cdot}341\dot{6}$ lbs.

XVI. Exercises. Pp. 161–2.

(1) $13\frac{1}{3}$ ozs. (2) $10\frac{1}{2}$ ozs. (3) $3\frac{3}{4}$ ozs.; $3\frac{3}{7}$ ozs.; $8\frac{8}{9}$ ozs.; 2 ozs. (4) $h : b :: 5 : 12$. (5) 5 ozs. (6) 12 ozs. (7) $(3 + 4\sqrt{2})$ ozs. (8) 10 : 7. (9) $1\frac{3}{7}$ ozs. (10) $1 : 2\sqrt{6}$. (11) 2 : 3; 27 inches. (12) 264 lbs. (13) 12 ozs. (14) 100. (15) 9 cwts. 1 qr. 20 lbs. (16) 1 cwt. 1 qr. 4 lbs. (17) $15\frac{5}{9}$ lbs.

Examination (Chapter VI.) P. 163.

(7) 90 lbs. (8) 30°. (9) $\frac{20}{\sqrt{221}} = 1{\cdot}34$ lbs. nearly. (10) $\frac{23\sqrt{3}}{120}$ lbs. (11) 32 ozs. (12) $W = 4P$, and the weights of the moveable pulleys make no difference when

they are equal. (13) 35 lbs. (16) 8. (18) 3 lbs.; equation of work. (19) Second system, 5 pulleys in each block. (20) $\frac{1}{3}$. (21) A force equal to his weight. (22) 12 lbs. $7\frac{1}{2}$ ozs. It is a case of 4 moveable pulleys with a power equal to $3\frac{1}{2}$ ozs. (24) $17\frac{7}{9}$ lbs.

XVII. Exercises. P. 170.

(1) 4 ozs. (2) 12 ozs. in AB; 9 ozs. in BC; 5 ozs. in CD. (3) $3\frac{3}{4}$ ins. (4) 7 ins. or 1 in., as they act in the same or in opposite directions.

XVIII. Exercises. Pp. 180–1.

(1) $2\sqrt{31}$ lbs. (2) $P\sqrt{(5+2\sqrt{2})}$. (3) $P\sqrt{10}$. (4) $R=P\sqrt{5}$, and acts at middle point of BC. (5) $2\sqrt{61}$ lbs. (6) 17 lbs. (7) 1·2 lbs.; 1·6 lbs. (8) 2 : 1. (9) $\sqrt{21}$. (10) 400 lbs. (11) 10 ozs. (12) 4 lbs. (13) $R=2P$. (14) $6\sqrt{3}$. (15) 200 lbs.

XIX. Exercises. Pp. 189–91.

(1) No. (3) CB. (4) $\frac{10}{\sqrt{3}}$ ozs. (5) $10\sqrt{55+28\sqrt{3}}$. (6) 38·4; 28·8 lbs. (8) $6P$ where P is the force along AB. (9) AD. (11) 12 lbs.; 16 lbs. (12) 9 ozs.; no weight; 11·1 ozs. nearly; 15 ozs.

XX. Exercises. Pp. 198–200.

(1) $(1{\cdot}58)P$. (2) 10·2 lbs. (4) $\sqrt{97+15\sqrt{6}-39\sqrt{2}}$. (5) $(1{\cdot}67)P$. (6) $X=\sqrt{2}$; $Y=\sqrt{2}+2\sqrt{3}-6$. (8) 17·32 lbs. (9) 5·95; 4·91. (10) 6·928. (12) $P\sqrt{2}$. (14) 52 lbs. (15) 122·8 lbs. (16) § 133, Ex. (2). (17) $\frac{10\sqrt{2}}{1+\sqrt{3}}$; $\frac{20}{1+\sqrt{3}}$. (18) 24 lbs. nearly.

XXI. Exercises. Pp. 215–7.

(1) 2 : 3. (2) $70\frac{7}{12}$, $50\frac{5}{12}$. (3) 20 ins. (4) $6\frac{3}{7}$ ins. from one end. (5) $\frac{2}{3}$ up median line from angular point. (6) $2\sqrt{2}\,P$ along the diagonal. (7) 40 ins. from greater force. (8) 12 ins. (9) $4\frac{7}{9}$ ins. (11) $\frac{5}{8}$ rad. (12) $6\frac{1}{4}$ lbs.; $3\frac{3}{4}$ lbs. (13) $19\frac{1}{3}$ ins. from one end. (14) $2\frac{1}{2}$ lbs. (15) $1\frac{3}{4}$ lbs. (16) $\frac{5}{13}$ of middle line from one of the sides. (17) $7\frac{13}{17}$ ins. from 3 lbs. weight. (18) 40 lbs.; 24 lbs. (19) 16 lbs.; $\frac{7}{16}$ of the length from the 7 lbs. (20) As $\frac{3}{4} : \frac{1}{2} : \frac{1}{3} : \frac{5}{12}$.

XXII. Exercises. Pp. 232–3.

(1) $1 : \sqrt{3}$. (2) $\frac{W}{\sqrt{2}}$. (3) 1·2 lbs. (4) $\frac{4}{5}\sqrt{3}$ lbs. (5) $(8-3\sqrt{6}) \times 9{\cdot}6$ ins. from one end. (6) $2\sqrt{3} : 1$. (7) $6\sqrt{2}$ ozs. (8) 5 ins. (9) 5 ins. from end where weight acts. (10) 6 lbs. (11) 12 ins.; 27 lbs.

Examination (Chapter VII.) Pp. 240–7.

(2) 40 ins.; 20 ins. (3) $2\sqrt{31}$ lbs. (4) 6 lbs. (8) $3\frac{1}{3} : 3\frac{1}{3} : 1\frac{2}{3} : 1\frac{2}{3}$. (10) 10 lbs. (11) 1·74 ft. (12) $\frac{72}{13}(4-\sqrt{3})$ ins. from 4 lbs. end. (15) $4\sqrt{2-\sqrt{3}}$ lbs. (17) $8\sqrt{3}$; $16\sqrt{3}$ lbs. (18) $10\sqrt{3}$. (24) $\sqrt{201}$; $\sqrt{146}$; $\sqrt{91}$. (25) (*a*) 50 lbs. 25 ins.; 40 lbs. 7·5 ins. (*b*) 62 lbs., 31 ins.; 72 lbs., 13·5 ins. (26) § 126. (27) 11·25; 27·4. (28) 15 lbs., 13 lbs. (29) § 155. (30) 27·73. (35) § 128. (38) A pushing force at 60° with the plane, or vertically upwards; $R = 6\sqrt{3}$ or O. (39) $6\frac{7}{13}$ ft. from one end. (42) 4 : 9. (43) $\frac{1}{2}$ in. from one end. (44) 2·7; $1{\cdot}35\sqrt{3}$. (46) § 154. (47) If D be the middle point of AB, C the point in the string to which the weight is attached, and CE the perpendi-

cular on AB in its position of equilibrium $\frac{CE}{DE}=\frac{24}{7}$; and when AB is horizontal, the pivot is 6 ins. from D. (48) $(5+\sqrt{3})$; $(5-\sqrt{3})$ ozs. (49) $\sqrt{5}$ lbs., $2\sqrt{5}$ lbs. (51) 12 ins. (54) If C be the nail, B the end of stick against wall, $CB=\sqrt[3]{3}$ ft.; the pressure on the nail is $8\sqrt[3]{3}$ ozs., and that on the wall $8(\sqrt[3]{9}-1)$ ozs.

XXV. Exercises. Pp. 271-2.

(2) 240 ft. (3) $20(\sqrt{2}+1)=48{\cdot}28$ lbs. (4) $\frac{13}{30}$ of length from one end. (5) 45°. (6) $\sqrt{3}:1$. (7) 10 lbs. (8) $\frac{1}{9}$ of median line from centre. (9) $\frac{1}{4}$ diagonal from centre. (10) $\frac{1}{6}$ up median line from base. (11) 4 ft. from the other end. (13) If a = side of square, Centre of Gravity is $\frac{5a}{18}$ from centre of square. (14) 8 ins. up median line of larger $\triangle$ from common base. (15) $\frac{3}{17}$ of length of rod from middle point. (16) $3\sqrt{3}:5$.

XXVI. Exercises. P. 285.

(1) If A, B, be the middle points of the longer and shorter arms, C the angular point, and if $AD=\frac{1}{3}AB$, then the C of G is $\frac{1}{3}$ up DC. (2) $7\frac{9}{13}$ ins. from heavier weight. (3) $27\frac{306}{331}$ feet from the base. (4) $\frac{3a}{2(4+\sqrt{3})}$ from the common side, where a is the side of the square. (5) $3\frac{2}{3}$ ins. from base. (6) One-sixth of radius from centre of large circle. (7) $\frac{1}{24}$ of diagonal from centre.

Examination (Chapter VIII.) Pp. 285–7.

(2) If C and B be the middle points of the two parts and CB be bisected in D, D must be vertically under A, the point of suspension. The position of the wire may be found by drawing a horizontal line through A,

and marking off a distance AE equal to $\sqrt{\frac{3}{7}}$ of a, where a is length of half the wire, and then dropping a vertical $EF = \frac{2}{\sqrt{7}}a$. If FA be joined and an equilateral $\triangle$ described on it, the two sides will be the position of the wire. (4) If G be point required and GF be perpendicular to BC, then $CF = \frac{17}{30}a$, and $GF = \frac{7}{15}a$, where $a = CB$. (5) 3·464 ins. (8) Reckoning from corner opposite to that from which rectangle is taken $X = 3\frac{8}{17}$ ins., $Y = 2\frac{11}{34}$ ins. (9) If C be the centre, and A the point to which weight is fixed, mark off $AG = 1$ inch along AC, and G will be vertically under point of suspension. (10) 3·26 ins. from top. (11) $5\frac{26}{43}$ ins. from base of shorter. (12) 3·2 ins. from heavier weight. (16) 8 : 7. (18) $\frac{9}{28}$ of line joining 9 ozs. and 3 ozs. from the 9 ozs. weight.

THE END.

www.ingramcontent.com/pod-product-compliance
Lightning Source LLC
LaVergne TN
LVHW020218110826
845151LV00003B/754